Quantum Computing Compact

Bettina Just

Quantum Computing Compact

Spooky Action at a Distance
and Teleportation Easy
to Understand

 Springer

Bettina Just
THM Technische Hochschule Mittelhessen
Gießen, Germany

ISBN 978-3-662-65007-3   ISBN 978-3-662-65008-0   (eBook)
https://doi.org/10.1007/978-3-662-65008-0

This Springer imprint is published by the registered company Springer-Verlag GmbH, DE, part of Springer Nature.
The registered company address is: Heidelberger Platz 3, 14197 Berlin, Germany

For my whole wonderful family, and posthumously for Steffen Bohrmann.

Preface

- What is quantum entanglement?
- How do quantum algorithms work?

These are the two central questions answered in this book.

Quantum computing is talked about in times of artificial intelligence and fast algorithms. You can find articles not only in technical literature and popular science literature, but also in ordinary daily newspapers and journals. One hears about superfast computers that could break encryption systems, be used in material design and in the simulation of traffic flows and logistics, and speed up all applications of artificial intelligence enormously.

However, the subject is difficult to access. Anyone who wants to understand how the algorithms work and why they are so fast will find strange justifications in the popular science literature. There is talk of a quantum bit being both "0" and "1" at the same time, or of socks that are both red and blue, but at the same time monochrome. In the technical literature one quickly reads about unitary transformations in high-dimensional spaces, which are hidden in the formalism of quantum physics. Their meaning is not immediately obvious, even if one can calculate the formulas. This book aims to build a bridge here and provide an introduction to the topic of "quantum computing" that is scientifically correct but does not require any special prior knowledge of mathematics or physics.

Quantum computers (like conventional computers) consist of the hardware, that is the devices themselves, and the software, the algorithms that run on the devices. The smallest piece of hardware in a quantum computer is the quantum bit.

Quantum bits are tiny particles, about the size of electrons or photons (particles of light), which obey the laws of quantum physics. These laws contradict our classical mechanistic idea of the world. They seem so strange that the great physicist Niels Bohr reportedly said of them, "Those who are not shocked by quantum theory do not understand it."

The strangest phenomenon of quantum physics is the so-called quantum entanglement. Quantum particles seem to influence each other with superluminal speed over arbitrarily long distances. So a single quantum particle cannot be considered in isolation, but must always be seen as part of a system—if it is changed, the whole system changes. Quantum algorithms make use of this property.

The present book treats in its first part which experiments (in their very basic configuration) are done, which results they give, and why these results really contradict classical mechanics. For this purpose, a thought experiment of Einstein, Podolsky, and Rozen from 1935 is adapted to photons. All that needs to be accepted about photons (for the purposes of this book) is that they can have a direction of polarization—nothing more is needed.

The experimental-physical aspects are not dealt with, because it is "only" about the basic understanding of the experiments and their results.

It is then a long way from the single (quantum) bit to the (quantum) computer. Many bits have to be provided and coordinated. This is particularly difficult with quantum bits, because they are very sensitive. They change their value when they come into contact with any kind of matter. And their "coordination" includes the fact that they can be entangled with each other. Quantum computer hardware is currently the biggest difficulty in the use of quantum computers. A brief outline of quantum computer hardware can be found in the last chapter of the book.

The second part of the book then deals in detail—beyond the hardware—with the software, i.e., with the way quantum algorithms work. Circuits form the basis of algorithms, so quantum circuits are considered. Some quantum gates and the algorithm for teleportation, one of the most exciting and at the same time simplest algorithms in the world of quantum computing, are presented.

Special mathematical knowledge is not required for this, since all gates and the teleportation algorithm are shown graphically in this book. They are also calculated, but who does not like to calculate can understand all gates and the teleportation algorithm only by the illustrations. All formulas can be simply skipped.[1]

The book is aimed at readers who want to understand quantum computing in more detail, and build a simple, scientifically clean bridge between popular scientific accounts and the technical literature. No knowledge of mathematics, physics, or computer science is required. For the first part, however, the willingness to expand the classical mechanistic physical worldview is required. For the second part, the willingness to enter a whole new world of algorithms is needed. These are algorithms in which a change in one quantum bit potentially changes the behavior of all other quantum bits—systemic algorithms, so to speak.

After reading this book, you will know how quantum entanglement is detected and why Einstein called it "spooky action at a distance." You will understand the basic idea of how quantum algorithms work. You can delve further into the mathematical, physical, or computer science literature, depending on your area of expertise. And you can put popular science articles into your context.

This book originated from lectures held by the author at the University of Applied Sciences of Central Hesse (Technische Hochschule Mittelhessen—THM) and from numerous lectures on quantum computing inside and outside the THM. Thanks go to all listeners of these lectures and talks, without whose constructive and lively questions and suggestions the book would not have been possible at all.

---

[1]Quantum algorithms with, e.g., three qubits are usually described with unitary operations on a 2^3, i.e., 8-dimensional vector space. In this book, for the first time, a more descriptive representation on the 8 corners of a normal cube is chosen (a normal cube has eight corners).

Special thanks go to Janka Cholevas for patiently and dedicatedly creating the many graphics, to Philipp Rangel Martinez for the independently created and creative program for visualizing the cube, and to Jakob Czekansky for his tireless and universal technical assistance. I thank Helmut Roth and Günter Schilling for proofreading and their very helpful comments. Thanks also go to my colleagues Klaus Rinn, Andreas Dominik, Berthold Franzen, and Christoph Gallus, as well as to Antonia Just and to Carlo Trentanove for their great willingness to listen and to share their perspective on things with me. I would also like to thank Adrian Just, Yvonne Arnold, Ulrike Beckenkamp, and Martin Launert and all my friends for always being open for discussions. Special thanks go to Jochen Rau and Thomas Stahl for numerous meetings over coffee and dinner and for the sometimes hard-to-bear honest, but always very constructive and goal-oriented feedback.

Of course, thanks go to Martin Boerger and Sophia Leonhard from Springer Verlag for the suggestion to write the book in the first place and for their wonderful guidance on the way from idea to book. And an anticipated thanks goes to all readers for comments and suggestions—enjoy reading :).

Gießen, Germany   Bettina Just
October 2021

Contents

Introduction 1

A **classical bit** is (as is well known) an object that can take exactly one of two different states. These are usually called "0" and "1". Classical gates perform operations on bits and are the basis of classical computers.

For classical bits there are the most different physical realization possibilities. Usually it is a wire that does not carry current or carries current. But you could also imagine a lamp that lights up or doesn't light up. Or a pointer (comparable to the hand on a clock), which can assume the two positions "horizontal" or "vertical".

Classical bits have the following two properties, which are so self-evident that they are usually not even mentioned:

- (Realism for bits)
  The value of a bit is uniquely defined at any time of the calculation. It can be read out and the readout process does not change the value.
- (Locality for bits)
  Changing the value of a particular single bit does not change the value of any other bit.

© Springer-Verlag GmbH Germany, part of Springer Nature 2022
B. Just, *Quantum Computing Compact*,
https://doi.org/10.1007/978-3-662-65008-0_1

We take these two properties for granted because they belong to the foundations of classical mechanics:

- (Realism)
  Objects have properties such as their weight, color, speed, or size that can be measured and are not changed by the measurement.
- (Localism)
  An action at one point in space does not instantaneously affect physical objects at another point in space. The effect of the action at the first point must be transmitted to the other point via physical media, and this takes at least as long as it takes light to travel from the first to the second point.

Quantum bits, unlike classical bits, do not obey the laws of classical mechanics. They follow the laws of quantum mechanics, i.e. the physics of the very smallest particles.

A quantum bit is (roughly speaking) an object that can take the values "0" or "1" or anything BETWEEN. Quantum gates perform operations on qubits, and are the basis for quantum computers.

Quantum bits are currently realized mainly with photons, ions and with superconductivity. But one can also imagine a pointer that can assume the two positions "horizontal" or "vertical" - or any position BETWEEN them.

Quantum bits do not have the two properties of realism and localism of classical mechanics mentioned above. Instead, the following two properties apply to them, which are the essential basics for quantum computing:

- (Change in measurement)
  If a quantum bit is measured, it returns one of the two values "0" or "1", and never a value in between. It also assumes the measured value when it is measured. So if it had a value between "0" and "1" before the measurement, this is overwritten by the measurement.
- (Quantum entanglement)
  Change of a quantum bit at one point of space can instantaneously, i.e. in the same instant, change the properties of

another quantum bit. This change occurs faster than the light needs to travel the distance between the two quantum bits.

The fact that objects change their properties as a result of measurement does not occur in classical computer science, but is well known in the world around us. There are situations in which the measurement itself changes the situation, e.g. in the quality testing of components. If a load test is carried out here, the component is no longer as resilient afterwards as it was before. And anyone who has children knows anyway that they behave differently when they are being observed.

The phenomenon of quantum entanglement, however, is so amazing that it was called "spooky action at a distance" by Einstein, and Bohr reportedly said, "Anyone who is not horrified by quantum theory has not understood it." The properties of the smallest particles can change like the properties of an heir to the throne who becomes king the moment the old monarch dies. No matter how far away he is, he is instantaneously king - faster than it takes light to overcome the distance between him and the old monarch.

The first part of this book explains the phenomenon of quantum entanglement. What experiments have been done, and why do they allow no other conclusion than the "spooky action at a distance"?

The second part of the book then deals with quantum algorithms, i.e. algorithms on quantum bits. How can you imagine quantum algorithms, why are they so fast? The basic idea is that - because of quantum entanglement - a change on one qubit changes the whole system. An operation on one qubit can change all other qubits as well, if necessary. Examples of such operations on qubits, and the algorithm for teleportation as an illustrative example, are discussed in the second part of the book.

Part I

Quantum Entanglement

Photons as Qubits

2

2.1 The Discovery of Quantum Particles

At the beginning of the twentieth century, two important new theories were developed in physics that fundamentally changed the known world view: the theory of relativity and quantum physics.

The theory of relativity with its core message "nothing is faster than light" revolutionized the concept of space and time.

Quantum physics deals with the smallest particles. The original aim was to understand the properties of light with its refractions and reflections. It was about phenomena such as the reflection of a landscape in a lake or the play of colours on an oil puddle. Advances in laser technology made experiments possible here that simply could not be carried out before. The most famous of these is certainly the "double-slit experiment". It shows that light behaves like a wave or a particle, depending on the environment. "Quantum particles" were initially photons (i.e. light particles) and electrons.

Quantum mechanics was developed to explain the behaviour of light and other quantum particles. It is a mathematical formalism for physical systems. It makes predictions about what results experiments with quantum particle will yield. The predictions of

© Springer-Verlag GmbH Germany, part of Springer Nature 2022
B. Just, *Quantum Computing Compact*,
https://doi.org/10.1007/978-3-662-65008-0_2

quantum mechanics have since then been repeatedly confirmed in experiments, and never disproved.[1]

The behavior of quantum particles has been experimentally proven, but is not compatible with our mechanistic worldview. This postulates:

- Reality is "local": An action at one point in space does not have an instantaneous effect on another point in space. The effect in the other point occurs at the earliest after the time that the light needs to cover the distance between the two points.
- Reality is "real": Physical objects have properties, e.g. a mass, a temperature or a speed. If suitable measuring instruments are available, these properties can be measured. A correct measurement does not establish the property - the property already exists before the measurement.
- The objects to be measured do not determine which measurements we humans perform on them. Our free will determines which measurements we perform on the objects.

Experiments of quantum physics show that in the world of quantum particles, these three postulates cannot be fulfilled at the same time. At least one of them is violated. This has profound philosophical implications: The world is not as it seems to us classically. The behaviour of the smallest particles cannot be explained by our classical view of the world. And this is not

---

[1] Quantum mechanics is now a required course in any physics BSc program, but there are also a great many popular science books about it, for readers with little knowledge of mathematics or physics. Here are some recommendations: The books by Ripota [1] and Fischer [2] explain the development historically. They start from the research personalities and put their contribution into the overall context. The books by Zeilinger [3, 4] explain physics up to the twenty-first century in a very entertaining way, physically correct and without the need for special mathematical or physical prior knowledge. Feynman's "QED - (Quantum Electrodynamics)" [5] is a milestone in the popular scientific explanation of quantum physics. Those who have more mathematical and physical knowledge can enter the theory e. g. via Schwindt's "Tutorium Quantenmechanik" ("Quantum Mechanics Tutorial") [6].

because we do not know things, or they are too small to measure - these would still be possible explanations within the mechanistic world view.

According to their structure, the experiments presented in this
book were first presented in a paper published by Einstein,
Podolsky and Rozen [7] in 1935. In this paper, thought experiments
proposed to get to the heart of the paradox of quantum theory. The
paradox was later named the "EPR paradox" after the authors. It
states, very roughly, that quantum particles interact with each
other at superluminal speeds. The quantum particles can be e.g.
photons (light particles), electrons, ions or even atoms. At that
time, the experiments could not yet be carried out practically for
technical reasons. Over the decades, the technology kept
improving and the seemingly paradoxical behavior of quantum
particle was confirmed again and again, e.g. in 2015 for photons
by a large international research group [8], in 2017 for atoms at
LMU Munich [9] or in a spectacular experiment with 100,000
internet participants in 2018 [10]. For photons, it can already be
taught in schools as part of optical physics courses [11].

In the first part of this book, the experiments and their results
are presented for photons. The experimental-physical aspects,
such as the exact experimental setup or the avoidance of errone-
ous or inaccurate measurement results (that would be a separate
book), are not discussed however. The experiments are presented
idealized, i.e. with their basic setup and their experimental error-
corrected results. From now on, quantum particles are photons for
this book.

2.2   Properties of Photons

Laser technology has now matured to the point where a physical
apparatus can be used to produce single photons, i.e. single particles of light, and emit them along a given direction. For the purposes of this book, the apparatus is simply a box from which
individual photons, i.e. individual small particles, are emitted
along a beam direction on request, see Figs. 2.1 and 2.2.

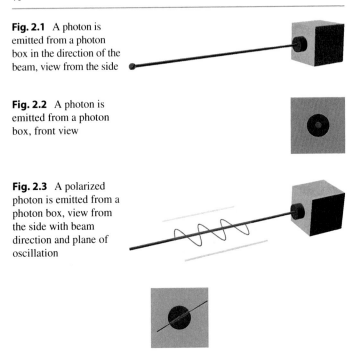

Fig. 2.1 A photon is emitted from a photon box in the direction of the beam, view from the side

Fig. 2.2 A photon is emitted from a photon box, front view

Fig. 2.3 A polarized photon is emitted from a photon box, view from the side with beam direction and plane of oscillation

Fig. 2.4 A polarized photon is emitted from a photon box, front view. One looks exactly along the plane, and therefore sees only a straight line - about the same as when one looks exactly from the front at a glass plate

A photon can (but does not have to) have a "polarization direction". It then oscillates in a plane on which the beam direction lies. The oscillation is a sine curve, the exact properties of which are of no interest for the purposes of this book. Only the plane itself is of interest. We can think of it as a glass plate in which the photon is trapped. Figures 2.3 and 2.4 show the situation when a polarized photon is emitted from the photon box.

Photons are so small that you cannot see them with the naked eye, or even under a microscope. To "see" them, you have to let them interact with an object. But the interaction can already change them - and every measurement is an interaction. So how

Fig. 2.5 Polarizing filter (front view)

Fig. 2.6 Angle of the polarizing filter (always in relation to the horizontal), here about 55°

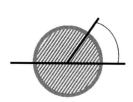

can you find properties if mere measurement changes the properties?[2]

You can send light - as photographers know—through a polarizing filter. For the purposes of this book, a polarizing filter is a round disc with a fine line pattern of parallel lines, see Fig. 2.5.

Depending on how the polarizing filter is rotated, the lines have an angle to the horizontal. This angle is called the "angle of the polarization filter", see Fig. 2.6.

A polarizing filter can be used to "measure" a photon emitted from a photon box. This measurement is made by holding the polarisation filter at a given angle perpendicular to the beam of the photon. Figures 2.7 and 2.8 show the situation for unpolarised photons from the side and from the front, respectively.

Figures 2.9 and 2.10 show the situation for polarized photons from the side and from the front, respectively.

---

[2]According to an untrue but beautiful story, Newton found the laws of gravity when he observed an apple falling from a tree. During the observation, photons were exchanged between Newton's retina and the apple, as they always are when a person sees something. This fact did not affect the fall of the apple. Imagine, however, that Newton's only observational option had been to throw other apples at the falling apple. This would have changed the fall of the observed apple. This is roughly how you can imagine the situation if you want to observe the smallest particles.

Fig. 2.7 Measurement
of a photon with a
polarization filter, view
from the side

Fig. 2.8 Measurement
of a photon with a
polarization filter, front
view

Fig. 2.9 Measurement
of a polarized photon
with a polarization filter,
view from the side

Fig. 2.10 Measurement of a polarized photon with a polarizing filter, front
view. Decisive for the measurement result is the difference angle a between
the polarization of the photon and the angle of the polarization filter

The measurement of a photon with a polarizing filter is sometimes compared to the situation of a Frisbee flying onto a manhole
cover. It always gives one of the following two results:

- Either the photon passes the polarization filter. Then it flies on,
  and is polarized in the direction of the polarization filter. If it
  had a different polarization than that of the filter before the
  measurement, or if it was not polarized, the measurement
  changes the polarization of the photon.

- Or the photon is absorbed. In this case, it does not pass through the polarization filter, but remains stuck in it, so to speak.

For the idealized experiments of this book, we assume that we can always decide unambiguously during the measurement whether the photon has passed or whether it has been absorbed. In fact, this can nowadays be achieved practically, with considerable effort for signal amplification.

Einstein assumed that there is no such thing as randomness in nature. "God does not throw dice" he is supposed to have said. He meant that, for example, even a coin toss with a normal coin contains no randomness. He assumed that one could calculate the result beforehand if one knew exactly the characteristics of the coin and exactly in which angle and with which speed it would hit. Of course, one could not predict it exactly. However, in Einstein's opinion, this would be due to the impossibility of describing and carrying out the experiment precisely enough, and not because the result is fundamentally undetermined. Accordingly, in his opinion, there are so-called "hidden variables" in the experiment, from which the result can be calculated unambiguously.

Quantum mechanics, which among other things describes the behavior of photons, postulates something different. In quantum mechanics, the result of measuring a photon is truly random (unless it is polarized exactly at the angle of the polarization filter). Quantum mechanics makes the following statements - confirmed again and again in experiments - about whether a photon will pass a polarization filter or will be absorbed:

- If the photon is not polarized, then for each angle the probability that it will pass or be absorbed is 1/2. Thus, if one measures many unpolarised photons in succession at the same angle or also at different angles, statistically half of them will pass or be absorbed in each case. The situation is like a coin about which nothing is known. Here, the best possible approach for modelling would also be 1/2 each for heads and for tails.
- If the photon is polarized, and α is the difference angle between the polarization angle of the photon and the angle of

the polarizing filter, then the photon passes with probability $(\cos \alpha)^2$.[3] This means in particular:

- If a polarized photon is measured in the direction of its polarization, it always passes (because $\cos 0 = 1$).
- If a polarized photon is measured perpendicular to its polarization, it never passes (because $\cos 90 = 0$).
- If a polarized photon is measured in a difference angle of $\alpha = 30°$ to its polarization, it will pass with probability 3/4 (because $\cos 30 = \sqrt{3} / 2$). So from a single photon one cannot predict whether it will pass or not. But if you do the experiment often, statistically 3/4 of the photons will pass.
- If a polarized photon is measured at a difference angle of $\alpha = 60°$ to its polarization, it will pass with probability 1/4 (because $\cos 60 = 1/2$). So again, for a single photon, one cannot predict whether it will pass or be absorbed, but one can predict that statistically for a large number of photons, a portion of 1/4 will pass, and a portion of 3/4 will be absorbed.

Experiments confirm these statistical predictions of the quantum mechanical calculation method. Malus' law from classical physics also states that statistically $(\cos \alpha)^2$ of the photons will pass when measured at the difference angle α.

But these experiments say nothing about whether it is already defined for each individual polarized photon whether it will pass BEFORE it hits the polarizing filter. It is possible that this is the case. But it is also possible that for each individual photon the measurement result is truly random, as the quantum mechanical interpretation says.

---

[3] If you do not have the cosine function present, and do not want to look it up in Wikipedia, all you need to know is: The cosine function assigns a number between −1 and 1 to each angle. The cosine of the right angle, i.e. of 90°, is 0, so its square is also 0. This corresponds to the picture that photons never pass if they are measured perpendicular to their polarization. The cosine of the angle of 0° is 1. The closer an angle is to horizontal, the larger the square of its cosine. This fits in the picture with the photons: The more similar the polarization of the photon and the angle at which it is measured, the greater the probability that the photon will pass.

2.3   The Experiments in this Book

This book explains what exactly is observed in the experiments when talking about the strange behavior of "entangled" photons.

The experiments described here are analogously the thought experiments proposed by Einstein, Podolsky and Rozen [7], which can nowadays also be carried out practically.

Bell [12] showed in 1964 that it is not possible to extend quantum mechanics with "hidden variables". That is, the world does not behave as predicted by the world view of classical mechanics. The derivation and explanation of this fact from the measurement results in the following book follows a special case of Bell's thought processes.

So it will be shown which idealized experiments are performed on pairs of photons to work out the strange behavior of quantum particles. "Idealized experiments" means that we consider photons as described above as objects coming out of a photon box, measured at a certain angle, and then passing and assuming the polarization of the angle, or being absorbed. The realization of the experiments is at most rudimentarily dealt with, and the treatment of measurement errors not at all - in idealized experiments there are no measurement errors.

For the experiments, pairs of photons are always considered in a way such that one is emitted to the left (as in the drawings above) and the other to the right out of a photon box. On the left, a person named Alice waits to measure the photon on the left. On the right, another person, Bob, waits to measure the photon on the right. We also call the photons "Alice's photon" and "Bob's photon" for short.

Figure 2.11 shows the situation.

In the first experiment, unentangled photon pairs are considered. These are obtained mentally by placing two photon boxes

Fig. 2.11 Set-up of the experiments. Photon pairs are emitted, one photon towards Alice and one towards Bob. Both then measure their respective photon

next to each other. Alice and Bob measure, and observe nothing surprising.

In the second experiment, entangled photon pairs are considered. Here, two photons each are emitted from a large photon box. These have been generated inside the box from one photon in a very specific way. Alice and Bob measure at the same angle, and their results are always the same for each pair. This really only allows us to conclude that there are hidden variables, although, as we shall see, there is also an absurd-sounding quantum mechanical interpretation that explains the behaviour.

The third experiment is again about entangled photon pairs, but now Alice and Bob no longer measure at the same angle. Instead, they very cleverly and randomly measure at angles of 0 degrees, 30 degrees, or 60 degrees (no other angles occur in this book). They are, as we shall see, statistically observing behavior that rules out the existence of hidden variables. However, it is fully explained by the absurd-sounding quantum mechanical interpretation.

Thus, if one wants to explain the results classically, the existence of hidden variables is simultaneously required and excluded – this is not possible, so there is no classical explanation. The quantum mechanical method of calculation perfectly predicts the statistical behavior. But it says that the world is not local, or not real, or that the photons and not the observer determine what the observer measures.

References

1. Peter Ripota. *Das Rätsel der Quanten (German Edition)*. Books on Demand, Dec 2016.
2. Ernst Peter Fischer. *Die Hintertreppe zum Quantensprung: Die Erforschung der kleinsten Teilchen von Max Planck bis Anton Zeilinger*. FISCHER Taschenbuch, Aug 2012.
3. Anton Zeilinger. *Einsteins Spuk: Teleportation und weitere Mysterien der Quantenphysik*. Goldmann Verlag, Jan 2007.
4. Anton Zeilinger. *Einsteins Schleier*. Goldmann Wilhelm GmbH, Mar 2020.
5. R. Feynman. *QED: Die seltsame Theorie des Lichts und der Materie*. Piper Verlag GmbH, May 2018.

6. Jan-Markus Schwindt. *Tutorium Quantenmechanik: von einem erfahrenen Tutor – für Physik- und Mathematikstudenten (German Edition)*. Springer Spektrum, Jun 2016.

7. A. Einstein, B. Podolsky, and N. Rosen. Can quantum-mechanical description of physical reality be considered complete? *Phys. Rev.*, 47:777–780, May 1935.

8. Lynden K. Shalm, Evan Meyer-Scott, Bradley G. Christensen, Peter Bierhorst, Michael A. Wayne, Martin J. Stevens, Thomas Gerrits, Scott Glancy, Deny R. Hamel, Michael S. Allman, Kevin J. Coakley, Shellee D. Dyer, Carson Hodge, Adriana E. Lita, Varun B. Verma, Camilla Lambrocco, Edward Tortorici, Alan L. Migdall, Yanbao Zhang, Daniel R. Kumor, William H. Farr, Francesco Marsili, Matthew D. Shaw, Jeffrey A. Stern, Carlos Abellán, Waldimar Amaya, Valerio Pruneri, Thomas Jennewein, Morgan W. Mitchell, Paul G. Kwiat, Joshua C. Bienfang, Richard P. Mirin, Emanuel Knill, and Sae Woo Nam. Strong loophole-free test of local realism. *Phys. Rev. Lett.*, 115:250402, Dec 2015.

9. Wenjamin Rosenfeld, Daniel Burchardt, Robert Garthoff, Kai Redeker, Norbert Ortegel, Markus Rau, and Harald Weinfurter. Event-ready bell test using entangled atoms simultaneously closing detection and locality loopholes. *Phys. Rev. Lett.*, 119:010402, Jul 2017.

10. Carlos Abellan and 106 other authors. Challenging local realism with human choices. *Nature*, 557(7704):212–216, May 2018.

11. Patrick Bronner. *Quantenoptische Experimente als Grundlage eines Curriculums zur Quantenphysik des Photons*. Logos Berlin, Berlin, 2010.

12. J. S. Bell. On the Einstein Podolsky Rosen paradox. *Physics Physique Fizika*, 1(3):195–200, November 1964.

The First Experiment: Independence

3

The photon boxes of the last chapter stand symbolically for an experimental setup in which single photons are emitted in the direction of a beam.

The photons can be polarized or non-polarized as described. Furthermore, they can be measured with a polarization filter. They then pass or are absorbed. When non-polarized photons are measured randomly, statistically half of the photons pass and half are absorbed. The first of the three experiments to illustrate quantum entanglement now uses two such photon filters. It is described in this chapter. Its result is not surprising.

3.1   Design of the Experiment

Two photon boxes are placed next to each other. They each emit a non-polarized photon at a time, synchronized: The left box emits with beam direction to the left, where Alice stands ready to measure it. The right box emits with beam direction to the right, where Bob is standing by (See Figs. 3.1 and 3.2)

For a photon pair, Alice measures first, then Bob. The polarization filters of Alice and Bob have the same angle (to the horizontal). This angle can be fixed beforehand. However, it can also be determined only when both photons have already emerged from

© Springer-Verlag GmbH Germany, part of Springer Nature 2022   19
B. Just, *Quantum Computing Compact*,
https://doi.org/10.1007/978-3-662-65008-0_3

Fig. 3.1 Setup of the first experiment. Non-polarized photon pairs are emitted, one photon in the direction of Alice and one in the direction of Bob

Fig. 3.2 View of the boxes from Alice and Bob respectively

View Alice   View Bob

Fig. 3.3 Measurement of Alice, then of Bob, at the same angle (shown here: 60°)

Measurement Alice   Measurement Bob

their boxes (Fig. 3.3). The question of interest in this experiment is not so much which photons exactly pass or are absorbed by Alice and by Bob. Rather, it is interesting to see at which **fraction of** the photon pairs Alice and Bob get the **same** result.

3.2   Result of the Experiment

It turns out that statistically half of the photons pass through and half are absorbed by both Alice and Bob. Moreover, the measurements of Alice and Bob seem to be independent. The four possible outcomes for a pair of photons are:

- the photon on Alice's side passes, so does the one on Bob's side;
- the photon on Alice's side passes, the one on Bob's side is absorbed;
- the photon on Alice's side is absorbed, the one on Bob's side passes;
- the photon on Alice's side is absorbed, the one on Bob's side is absorbed as well.

All four outcomes occur statistically with equal frequency.

So Alice and Bob get the **same** result (statistically) **half the time.**

3.3 Interpretation of the Result—Independence

The experiment is easy to interpret. The measurements of Alice and Bob are basically two independent coin tosses with a coin that indicates "passed" or "absorbed" with probability 1/2.

The Second Experiment: Equality

4

The second of the three experiments illustrating quantum entanglement looks like the first from the outside. However, the photon pairs are now not emitted from two independent boxes. They come from only one box, inside of which the photons are "entangled". Alice and Bob measure as before. The result of the experiment is completely different from that of the first experiment, but in itself also well explainable.

4.1   Design of the Experiment

Photon pairs are emitted from a specially prepared photon box—again, one to the left for Alice and one to the right for Bob, see Fig. 4.1.

The box looks from the outside like the box from the first experiment (see previous chapter). Inside, however, the two photons are not created independently of each other. Instead, they are created from a single photon that is sent through an "entanglement crystal". This splits it into two photons, each with half the energy, which are then emitted out of the box in the direction of Alice and Bob, see Fig. 4.2.

We will not go into the technical details here. For the purpose of this book it is sufficient to know that a production of such

© Springer-Verlag GmbH Germany, part of Springer Nature 2022   23
B. Just, *Quantum Computing Compact*,
https://doi.org/10.1007/978-3-662-65008-0_4

Alice_____■_____Bob

Fig. 4.1 Setup of the second experiment. Photon pairs are emitted from ONE box, one photon in the direction of Alice and one in the direction of Bob

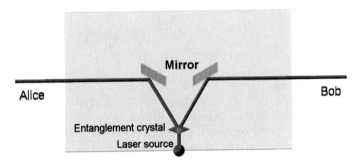

Fig. 4.2 View inside the box with entanglement crystal

Fig. 4.3 Measurement of Alice, then of Bob, at the same angle (shown here: 55°)

Measurement Alice   Measurement Bob

"entangled" photons is easily possible, and that they can then be measured by Alice and Bob as in the first experiment.

This means again: For a photon pair Alice measures first, then Bob. The polarization filters of Alice and Bob have the same angle (to the horizontal, see Fig. 4.3).[1] This angle can be fixed before-

---

[1]Technically savvy readers will notice: The book suppresses the fact, that physically Bob has to do a rotation of 90° relative to Alice. This is true and may be forgiven, as it does not change anything in the effects, but would make the presentation more complicated.

hand. However, it can also be determined when both photons have already left the box.

Again, we are mainly interested in the **fraction of** photon pairs where Alice and Bob get the **same** result.

4.2   Result of the Experiment

It turns out again that, statistically, half of the photons pass through and half are absorbed at both Alice and Bob.

But:
Alice and Bob get the same result in **all cases.**

4.3   Interpretation of the Result

4.3.1   Classical Interpretation: Hidden Variables

The perfect correlation observed in this experiment does not occur with independent coin tosses. The result is more reminiscent of the following experiment with "hidden variables" (a term coined by Albert Einstein):

Two balls at a time are dyed either both red or both blue, hidden from the observer. They are then individually wrapped. One of the balls is sent to Alice and one to Bob. For Alice and Bob, the probability is 1/2 each of finding a red or a blue ball. Alice opens her package first. She sees the color of her ball, and thus knows the color of Bob's ball.

In this experiment, true randomness exists, if at all, only at the very beginning when selecting the color of a pair of balls. After that, each pair of balls has its color (its "hidden variable"). The uncertainty in the experiment is thus due to incomplete knowledge on the part of Alice and Bob, and not to properties of physical reality.

So the classical interpretation of the result of the second experiment would be the following: For each photon pair, when it exits the box, it is already uniquely determined for each angle whether (both photons) will pass, or whether (both photons) will be absorbed. In this sense, the photon pairs have a hidden variable for each angle. These may differ for different photon pairs, but not for the two photons of a photon pair.

4.3.2 Quantum Mechanical Interpretation: System State, Instantaneous

In the quantum mechanical interpretation, it is undetermined until the measurement whether Alice's photon passes or is absorbed. In her measurement it passes with probability 1/2, and is polarized afterwards in the direction of her polarization filter. With probability 1/2 it is absorbed.

Bob's photon, according to quantum mechanical interpretation, takes the **state of Alice's photon at the moment of Alice's measurement**. That means, if Alice's photon passes, it is instantaneously, i.e. without time delay, also polarized in this direction. It then also passes, because it is measured by Bob in the direction of its polarization. If Alice's photon is absorbed, Bob's photon takes a polarization perpendicular to the direction of measurement of the polarization filter, and is thus will be also absorbed in Bob's measurement.

This interpretation sounds completely absurd. Einstein called instantaneous information transfer "spooky action at a distance". For how should Alice's measurement cause the properties of Bob's photon to change without any loss of time (i.e. at faster-than-light speed)? Doesn't that contradict the theory of relativity? Well, the interpretation does not contradict the theory of relativity. Because theory of relativity only says that no material can move through space faster than light. And in the experiment, no material is sent between Alice's and Bob's photon—just information is transported faster than the speed of light between them.

The quantum mechanical interpretation also states that entangled photons must be considered as a system, independent of the distance in the underlying space. A change of one object of the system instantaneously changes the state of the other object, also if the objects are distant in space.

Since the quantum mechanical interpretation is very strange, a third experiment is done to see which interpretation is more viable.

The Third Experiment: Spooky Action at a Distance

5

The third experiment is actually done to disprove the quantum mechanical interpretation of the second. It is supposed to show that there is no "spooky action at a distance". But it confirms the quantum mechanical prediction and shows instead that there can be no hidden variables in the passing or absorption at polarization filters of entangled photon pairs.

5.1 Setup of the Experiment

As in the second experiment, pairs of entangled photons are sent to the left to Alice and to the right to Bob, cf. Fig. 5.1.

Alice decides randomly and with probability 1/2 for each photon pair whether she measures at 0° or 30° (to the horizontal). Bob also decides randomly and with probability 1/2, and independently of Alice, whether he measures at 30° or at 60°. Again, both can also decide which angle to measure after the photon pair has

© Springer-Verlag GmbH Germany, part of Springer Nature 2022
B. Just, *Quantum Computing Compact*,
https://doi.org/10.1007/978-3-662-65008-0_5

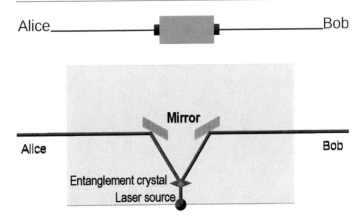

Fig. 5.1 The emission of entangled photons is identical in the second and third experiments. Both experiments differ in the way Alice and Bob measure

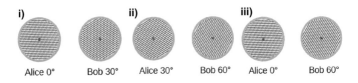

Fig. 5.2 Possible measurements in the third experiment

departed, and again Alice measures first.[1] Bob, however, can measure immediately afterwards—faster than it would take the light to get from Alice to him. Results where both measured at 30° are erased. This is because measurements at the same angle have already been analysed in the second experiment.

Possible measurements are therefore, see Fig. 5.2:

(a) Alice measures at 0°, Bob at 30°;
(b) Alice measures at 30°, Bob at 60°;
(c) Alice measures at 0°, Bob at 60°.

---

[1] Someone has to measure first—a truly exactly simultaneous measurement is not experimentally possible.

5.2 Result of the Experiment

In the measurements of both Alice and Bob, statistically half of the photons again pass, and half are absorbed. This is also the case if one considers only the measurements with respect to one of the angles occurring at Alice or Bob.

We now consider again at what **fraction** of the photon pairs of Alice and Bob get the same result. This is distinguished according to how Alice and Bob measured. It turns out:

- The portion of **equal** results is statistically **3/4** if Alice measures at 0° and Bob measures at 30°;
- The portion of **equal** results is statistically **3/4** if Alice measures at 30° and Bob measures at 60°;
- The portion of **different** outcomes is statistically **3/4** if Alice measures at 0° and Bob measures at 60°.

5.3 Interpretation of the Result

As described, if the difference angle of Alice's measurement and Bob's measurement is 30°, the portion of equal outcomes is (statistically) 3/4. If the difference angle of Alice's measurement and Bob's measurement is 60°, the portion of different outcomes is (statistically) 3/4.

It now turns out that this is incompatible with the existence of hidden variables, but is perfectly consistent with the quantum mechanical interpretation.

5.3.1 Classical Interpretation: Hidden Variables

In this interpretation, it is assumed that for each photon pair there are hidden variables for the measurements at 0°, 30° and 60°. That is, for each photon pair, it is already determined at emission whether both will pass or be absorbed when measured with a polarizing filter in the 0°, 30°, and 60° directions. That both will always pass or be absorbed is the result of the second experiment.

Table 5.1 Possibilities of hidden variables for 0°, 30°, 60°

0°	30°	60°
Absorbs	Absorbs	Absorbs
Absorbs	Absorbs	Passes
Absorbs	Passes	Absorbs
Absorbs	Passes	Passes
Passes	Absorbs	Absorbs
Passes	Absorbs	Passes
Passes	Passes	Absorbs
Passes	Passes	Passes

There are theoretically eight possibilities how the hidden variables for a photon pair can look like for the three angles, see Table 5.1. These possibilities still have nothing to do with the actually performed measurement. They only describe the possible values of the hidden variables.

For each assignment of the hidden variable, it is unambiguously determined whether or not Alice and Bob get the same results when they are measured.

For example, for the fourth option (absorb at 0°, pass at 30 and 60°), Alice and Bob get different results if Alice measures at 0° and Bob measures at 30° or at 60°. They get the same result if Alice measures at 30° and Bob at 60°.

The difference angle between Alice's measurement and Bob's is always 30° or 60°. No other difference angles occur. The central idea of John Bell [1] in 1964 (when Einstein was already dead) was to cleverly combine the following two questions:

- Do Alice and Bob get the same result when the difference in their measurement angles is 30° (that is, when the difference angle is small)?
- Do Alice and Bob get a different result if the difference in their measurement angles is 60° (that is, if the difference angle is large)?

Depending on the assignment of the hidden variables, the results are different. They are listed in Table 5.2.

Table 5.2 Measurement results with given hidden variables

0°	30°	60°	Alice 0 Bob 30 results the same?	Alice 30 Bob 60 results the same?	Alice 0 Bob 60 results different?
Absorbs	Absorbs	Absorbs	Yes	Yes	No
Absorbs	Absorbs	Passes	Yes	No	Yes
Absorbs	Passes	Absorbs	No	No	No
Absorbs	Passes	Passes	No	Yes	Yes
Passes	Absorbs	Absorbs	No	Yes	Yes
Passes	Absorbs	Passes	No	No	No
Passes	Passes	Absorbs	Yes	No	Yes
Passes	Passes	Passes	Yes	Yes	No

The two sub-questions are combined to form the following question:

Question "If Alice measured at 0° and Bob measured at 30°, are the results of both measurements the same? Otherwise, if Alice measured at 30° and Bob measured at 60°, are the results of the two measurements the same? Otherwise, so if Alice measured at 0° and Bob measured at 60°, are the results of the two measurements different?"

This question is statistically answered with "yes" in 3/4 of all cases by the measurement results actually obtained, as described in the previous Sect. 5.2. Because there it was observed:

- The proportion of **equal** results is statistically **3/4** if Alice measures at 0° and Bob measures at 30°;
- The proportion of **equal** results is statistically **3/4** when Alice measures at 30° and Bob measures at 60°;

- The proportion of **different** outcomes is statistically **3/4** if Alice measures at 0° and Bob measures at 60°.

But the table of measurement results given hidden variables says otherwise: if the hidden variables are fixed, and the measurement of Alice and Bob is fixed independently, the probability of answering "yes" to the question is at most 2/3 in all cases. This makes it impossible to get an (arbitrarily weighted) statistical mean of ¾ as observed in the third experiment.

But if there are no hidden variables—how is the result of the second experiment possible? In a moment, we will see how quantum mechanics can explain it.

5.3.2   Quantum Mechanical Interpretation: System State, Instantaneous

Once again as a reminder, as already described in the second experiment: In the quantum mechanical interpretation, whether Alice's photon passes or is absorbed is undetermined until the measurement. In her measurement, it passes with probability 1/2, and is polarized afterwards in the direction of her polarization filter. With probability 1/2 it is absorbed.

According to quantum mechanical interpretation, Bob's photon assumes the **state of Alice's photon at the moment of Alice's measurement**. More precisely: If Alice's photon passes, Bob's photon is instantaneously, i.e. without time delay, also polarized in this direction. If Alice's photon is absorbed, Bob's photon assumes polarization perpendicular to the measurement direction of Alice's polarizing filter. The possible states of Alice's and Bob's measurements are illustrated in Fig. 5.3.

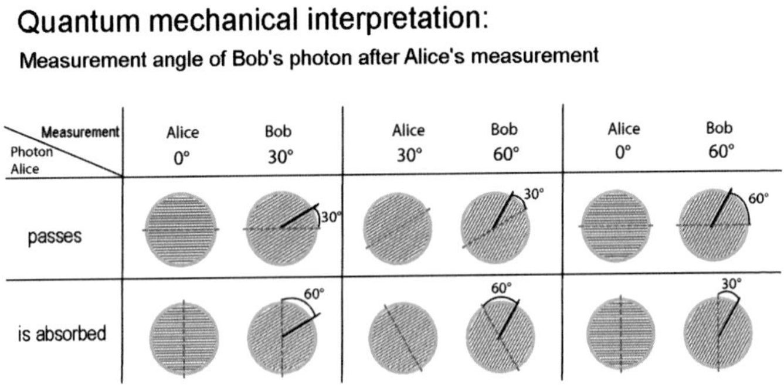

Quantum mechanical interpretation:
Measurement angle of Bob's photon after Alice's measurement

Fig. 5.3 Possible states after Alice's measurement in the quantum mechanical interpretation

We now show that the quantum mechanical interpretation agrees perfectly with the results of the third experiment. Because:

- If Alice's photon has passed her polarizing filter, then according to the quantum mechanical interpretation Bob's photon is polarized in the direction of her polarizing filter.

  If Bob's photon is then measured at a difference angle of $30°$, the probability that it will also pass is according to the (classical) law of Malus $(\cos 30)^2 = 3/4$.

  If Bob's photon is measured at a difference angle of $60°$, the probability that it will pass is according to the law of Malus $(\cos 60)^2 = 1/4$. The probability that it will not pass (and that is what is being asked) is therefore $3/4$.

- If Alice's photon has not passed her polarizing filter, then according to the quantum mechanical interpretation Bob's photon is polarized in the direction perpendicular to Alice's polarizing filter.

  If Bob's photon is then measured at a difference angle of $30°$ to Alice's measurement, that is a measurement at a difference angle of $60°$ to his polarization direction. The probability that it passes is according to law of Malus $(\cos 60)^2 = 1/4$. The probability that it doesn't pass (and that is what is being asked—whether it does not pass as well as Alice's photon did not pass) is therefore $3/4$.

  If Bob's photon is measured at a difference angle of $60°$ from Alice's measurement, it is a measurement at an angle of $30°$ to its polarization. The probability that it will pass (and thus the measurement of Bob's photon will differ from the measurement of Alice's photon) is according to Malus' law $(\cos 30)^2 = 3/4$.

The quantum mechanical interpretation coincides perfectly with the observed results. But how can it be that the measurement on Alice's photon instantaneously affects the properties of Bob's photon? This contradicts our classical worldview. But it has been experimentally demonstrated that any "information" whatsoever that travels from Alice's photon to Bob's photon travels at least

10,000 times the speed of light.[2] This is the reason for the saying attributed to Bohr, "For if one is not shocked by quantum theory when first coming around with it, one cannot possibly have understood it."

References

1. J. S. Bell. On the Einstein Podolsky Rosen paradox. *Physics Physique Fizika*, 1(3):195–200, November 1964.
2. Juan Yin, Yuan Cao, Hai-Lin Yong, Ji-Gang Ren, Hao Liang, ShengKai Liao, Fei Zhou, Chang Liu, Yu-Ping Wu, Ge-Sheng Pan, et al. Lower bound on the speed of nonlocal correlations without locality and measurement choice loopholes. *Physical Review Letters*, 110(26), Jun 2013.

---

[2]There are many experiments on this, e.g., [2].

Evaluations and Interpretations

6

6.1 Structural Observations in the Experiments

The three experiments from the last chapters have provided the following observations about the properties of entangled photons[1]:

- When measured both at any identical angle, also randomly determined by the observers, both photons always behave the same, i.e. both pass or both are absorbed. This is also true if the angle is determined only after both photons have already left the photon source. (Result of the second experiment)
- There can be no "hidden variables" of the photon pairs, which would define the result of the measurement before carrying out the measurement. Thus it is not the case, that the result is defined—but yet unknown to the observer—before observing it. (Analysis of the third experiment.)
- If any "information" about the measurement result should be transferred from one photon to the other, this is done with at least 10,000 times the speed of light. (Proven by experimental physics, see e.g. [7].).

---

[1] It was always assumed that the observer can perform his measurements independently of the observed photons. This is not the case if everything is predetermined, and both the experiments performed and their results are already fixed in a sense that the observer cannot influence. Some believe this, but it would contradict the free will of the observer.

© Springer-Verlag GmbH Germany, part of Springer Nature 2022
B. Just, *Quantum Computing Compact*,
https://doi.org/10.1007/978-3-662-65008-0_6

- The same experimental setup does not always deliver the same measurement result in the world of photons (first experiment). These are not errors in the experimental setup, but the measurement results actually arise only through the measurement, and are not already defined by the experimental setup.

6.2 Modeling in Quantum Theory and Philosophical Implications

Quantum theory is a mathematical model. It predicts the experimental results of experiments statistically correctly. It includes the following model assumptions:

- Non-realism: There is real randomness. The same experimental set-up does not always produce the same result. This is modelled in quantum physics with the help of probability theory. The probability distribution on which randomness is based is the so-called "wave function". It assigns a probability to each measurement result, and can then be statistically confirmed by experiments. If measurements are actually made, the possible results no longer have probabilities, but one has occurred and the others have not. One then speaks of the "collapse of the wave function" during the measurement.

    In the example of measuring a horizontally polarized photon at a 45-degree angle, the wave function is that of a coin toss: With probability 1/2 the photon passes, with probability 1/2 it is absorbed.

    Thus, a quantum particle does not have a property per se, which is then detected by the measurement. Rather, it is only through the measurement that the particle acquires a property. In this sense, quantum particles behave like a voter who is still thinking about whom to vote for on the way to the voting booth. Only by the measurement, i.e. by the cross on the ballot paper, the voter gets the property to be a voter of a certain candidate. Before, he was "at the same time" a voter of all candidates, which come into consideration for him.

- Non-locality: In entangled quantum pairs, the measurement of one quantum particle instantaneously affects the state of the other, i.e., at the same instant. The change of state thus takes place with superluminal speed.

Both model assumptions contradict the assumptions of classical mechanics.

- Classical mechanics assumes that one can unambiguously calculate the result of an experiment if one knows all the necessary input variables. In this sense, a dice throw is not random. If one knew the exact forces applied during the throw, as well as the material properties of the dice and the table, the properties of the throwing hand, etc., one could—according to classical mechanics—calculate what the result would be.

  This is because in classical mechanics, objects have properties, such as their weight, their position in space, or their color. Correct experimental setups detect these properties, and the experiments can be repeated and always give the same results. According to the rules of classical mechanics, inanimate objects do not behave like still uncertain voters.
- Classical mechanics assumes that experiments are carried out in a spatially closed area in a "closed system". If one subsystem influences another, this takes time. The impact of a billiard ball, for example, only influences the cushion after the billiard ball has arrived there.

The bizarre-seeming assumptions of quantum physics have been confirmed experimentally again and again, and never disproved. They do, however, raise some philosophical questions:

- Is there real randomness in the inanimate world? Einstein still said "God does not play dice". However, he died already in 1955, and thus could not know the considerations of Bell (1964).
- The old idea of the outside observer, who detects an objective reality independent of him, is dismissed. The observer becomes

part of the system he observes and influences it through his observation.[2]

- The idea of space and the flow of information in space is called into question. According to the theory of relativity, matter cannot move through space at the speed of faster than light. But in the case of entangled pairs of particles, the measurement of one particle instantaneously affects the state of the other particle. Matter is not moved in this process, so it does not contradict relativity theory. But how does this information flow?

Within the scope of this book, the philosophical aspects cannot be dealt with in detail. Therefore, here are recommendations for further reading:

- For the physics of inanimate matter, the most widely accepted conception is the "Copenhagen interpretation". It was developed in 1927 by Heisenberg and Bohr, and since then it has been repeatedly illuminated and differentiated. A nice overview article can be found in Wikipedia, there are also numerous YouTube videos about it.
- Heisenberg wrote some papers on the connection between (quantum) physics and philosophy, see [4, 5].
- Bohr and Einstein had different opinions about the role of randomness in physics. They had the so-called "Bohr-Einstein debate" about it, see [6].
- About physics, spirituality and transcendence all great physicists have thought at the beginning of the twentieth century in the face of quantum physics and relativity. An overview can be found, for example, in [3]. A connection between quantum physics and Hinduism/Buddhism is explicitly drawn by Capra [2].
- The mathematics of quantum mechanics is now also used to describe human behaviour. Imagine a married couple who want to buy a car. Both are still undecided whether it will be

---

[2]An almost human characteristic of the system: people also behave differently depending on whether they believe they are being observed or not.

brand A or brand B, but both are sure that they will agree. The situation is similar to that of a pair of entangled photons—the outcome is uncertain, but it will be the same for both. An introduction to quantum physical models in describing human behavior and decision making can be found in [1].

References

1. Jerome Busemeyer and Peter Bruza. *Quantum models of cognition and decision*. Cambridge University Press, Cambridge, 2012.
2. Fritjof Capra. *The Tao of Physics (Flamingo)*. Flamingo Harpercollins, Mar 1992.
3. Hans P. Dürr. *Physik und Transzendenz die großen Physiker unserer Zeit über ihre Begegnung mit dem Wunderbaren*. Driediger, Bad Essen, 2010.
4. Werner Heisenberg. *Quantentheorie und Philosophie : Vorlesungen u. Aufsätze*. Reclam, Stuttgart, 1979.
5. Werner Heisenberg. *Physik und Philosophie*. Hirzel S. Verlag, Aug 2011.
6. Manjit Kumar. *Quanten : Einstein, Bohr und die große Debatte über das Wesen der Wirklichkeit*. Berlin Verlag, Berlin, 2009.
7. Juan Yin, Yuan Cao, Hai-Lin Yong, Ji-Gang Ren, Hao Liang, Sheng-Kai Liao, Fei Zhou, Chang Liu, Yu-Ping Wu, Ge-Sheng Pan, and et al. Lower bound on the speed of nonlocal correlations without locality and measurement choice loopholes. *Physical Review Letters*, 110(26), Jun 2013.

Part II

Quantum Computing with the Example of Teleportation

Quantum Algorithms Vividly

7

In classical computer science, a bit is an object that can take the two values "0" or "1". The values are sometimes also referred to as "false" and "true". In classical computer science, the following two facts are so clear that they are usually not even explicitly stated:

- (Realism)
  The value of a bit is uniquely defined at any time of the calculation. It can be read out and the readout process does not change the value.
- (Locality)
  Changing the value of a particular single bit does not change the value of any other bit instantaneously.

In quantum computing, a qubit is an object that can take the values "0" or "1" *or anything in between.* The two boundary values are usually denoted by "$|0\rangle$" and "$|1\rangle$". At first, these are just strange looking labels. Photons, as considered in the first part of the book, are typical examples of qubits. The polarization of a photon is "horizontal" (corresponding to "$|0\rangle$") or "vertical" (corresponding to "$|1\rangle$"), or *anything in between*—any polarization angle is possible.

When working with qubits, however, the assumptions of realism and locality that are so self-evident for classical bits must be

© Springer-Verlag GmbH Germany, part of Springer Nature 2022   45
B. Just, *Quantum Computing Compact*,
https://doi.org/10.1007/978-3-662-65008-0_7

partially abandoned. The reasons for this were explained for photons in the first part of the book:

- (Restriction of realism)
  Measuring generally changes the polarization of a photon.
  If, for example, a photon polarized at an angle of 45° is measured with a horizontal polarization filter, there are two possible measurement results: Either the photon passes the filter, and is subsequently horizontally polarized itself, or it is absorbed by the filter. After the measurement, it is therefore in any case no longer polarized at a 45° angle.

  And: If we do not know the polarization of the photon before the measurement, we cannot find it out by a measurement. Because measuring changes the polarization of the photon.

- (Restriction of locality)
  Changing the value of a particular single bit may change the value of one (or even more) other bits. In the second experiment in the first part of the book, we saw that measuring a photon fixes the measured value of a photon entangled with it, even if (as shown in the third experiment) it was not fixed before.

How can one then imagine calculations with such qubits? At first, a graphical illustration follows to get a very first idea. Further details and a technically exact presentation are then the subject of the following chapters.

For three qubits we imagine a normal three-dimensional cube for the very first view. This has eight corners (see for yourself).

Further, we imagine that at each corner of the cube is a hollow glass sphere. All these spheres are the same size, they can each hold one unit of a liquid. The spheres are connected along the edges of the cube (and only there) with thin glass tubes through which the liquid can flow. However, it cannot stay there. The whole structure is reminiscent of the Atomium in Brussels (see Fig. 7.1).

A quantum algorithm is now a sequence of "shakes" of the cube. What exactly a "shake" is not interesting for the moment.

Fig. 7.1 A cube with
glass spheres at the
corners and glass tubes
at the edges—for the
very first visualization of
a quantum algorithm

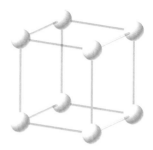

We simply imagine that one can "shake" in the three directions of
space (i.e., left-right, bottom-up, and front-back). Shaking in a
particular direction causes fluid to spread along the glass tubes in
that direction (and only along those glass tubes) to the adjacent
two spheres.

Here comes a concrete task: The lower left front glass sphere is
completely filled, all other glass spheres are empty. How often do
you have to shake so that the liquid is evenly distributed in all
eight spheres?

Figure 7.2 shows a method that requires only three shakes.

You can see: Three steps are enough to change eight values. A
classical algorithm needs at least eight steps to change eight values, because it has to access every single value at least once. This
is the first glimpse of why quantum algorithms are so fast.

To describe more exactly what quantum algorithms do, the
very first idea with the spheres has to be adjusted a bit. Instead
of spheres, we need to imagine that there are squares with side
length 1 (and therefore also with area 1) attached to the corners. So the algorithm above looks as shown in Fig. 7.3.

Note that the side length of the (green) colored square in the
start state is 1. After the first shake, the two colored squares each
have side length $1/\sqrt{2}$, so each has area 1/2. The second shake
yields four small squares with side length 1/2, so each has area
1/4, and in the final state there are eight small squares with side
length $1/\sqrt{8}$, so each has area 1/8.

If you add up the areas of all the small squares, you always get
the original total area of 1.

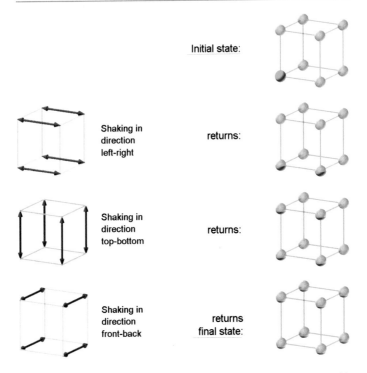

Initial state:

Shaking in direction left-right

returns:

Shaking in direction top-bottom

returns:

Shaking in direction front-back

returns final state:

Fig. 7.2 Illustration of a simple quantum algorithm on three qubits—with spheres for a first idea. Shaking three times distributes the liquid from one sphere evenly to all eight spheres

In the following chapters of this part of the book, we will show what this illustration has to do with quantum bits (in short: "qubits"), and what the "shaking" exactly means.

Just this much in advance:

- Each qubit is responsible for a certain spatial direction. The first qubit for the direction left-right, the second for the direction down-up, and the third for the direction front-back.
- Each of the eight corners of the cube represents a possible measurement result that can come out if you measure all three qubits.

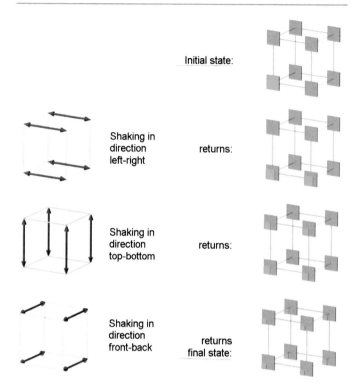

Initial state:

Shaking in direction left-right   returns:

Shaking in direction top-bottom   returns:

Shaking in direction front-back   returns final state:

Fig. 7.3 Illustration of the quantum algorithm on three qubits—with squares, the correct representation

The first qubit can provide the measurement result $|0\rangle$ or $|1\rangle$, as can the second and third qubit. The possible measurement results are therefore.

$|0\rangle\,|0\rangle\,|0\rangle$, $|0\rangle\,|0\rangle\,|1\rangle$, $|0\rangle\,|1\rangle\,|0\rangle$, $|0\rangle\,|1\rangle\,|1\rangle$, $|1\rangle\,|0\rangle\,|1\rangle$, $|1\rangle\,|0\rangle$ $|1\rangle$, $|1\rangle\,|1\rangle\,|0\rangle$ or $|1\rangle|1\rangle\,|1\rangle$. Shorter one writes for it, simply to get rid of some vertical lines and square brackets:

$|000\rangle$, $|001\rangle$, $|010\rangle$, $|011\rangle$, $|100\rangle$, $|101\rangle$, $|110\rangle$ and $|111\rangle$.

- The area sum of the 8 small colored squares is 1 at any time. It will turn out that the areas of the small squares are probabilities: The area of the small square at a possible measurement result is the probability that measuring all three qubits will produce exactly that measurement result.

- Changes to a single one of the three qubits can cause all eight probabilities to change. This is the case, for example, in the last step of the shaking algorithm shown above. Here, shaking once along the third qubit, changes all the probabilities.

Quantum Bits and Quantum Registers

8

8.1   Representation of a Qubit for Algorithms

A qubit $|q\rangle$ is defined in quantum computing as an object of the shape

$$|q\rangle = a \cdot |0\rangle + b \cdot |1\rangle,$$

that can be "measured."[1]

Here $|0\rangle$ and $|1\rangle$ are the so-called "basis states". a and b are real numbers between -1 and $+1$ in this book, they are called "probability amplitudes".[2] It must be true that $a^2 + b^2 = 1$. Measuring the qubit yields one of the basis states, namely the basis state $|0\rangle$ with probability a^2, and the basis state $|1\rangle$ with probability b^2.

---

[1] It is strange that here a bracket is opened with "|" and closed again with ")". But this is the common notation for qubits. It is Dirac's bra-ket notation, which refers to the dot product $\langle\, .\, |\, .\, \rangle$ The "ket vector" $|\, .\, \rangle$ is the right-hand part of the dot product.

[2] The restriction to real numbers is a simplification for the elementary introduction in this book. In the literature, a and b are complex numbers whose absolute squares add up to 1. For readers who are familiar with complex numbers, there is an explanation of this at the end of the chapter.

© Springer-Verlag GmbH Germany, part of Springer Nature 2022   51
B. Just, *Quantum Computing Compact*,
https://doi.org/10.1007/978-3-662-65008-0_8

A light particle polarized at an angle α to the horizontal is a possible physical realization of such a qubit.[3] With polarized photons it is easy to see why the probabilities are a^2 and b^2, and not a and b. Figure 8.1 shows the situation with the example angles 30° and 120°.

If you want to look closely (and remember a little bit about sine and cosine), you will find here in the figure with the 30° angle a right triangle with the side lengths cos 30°, sin 30° and 1 (since the circle has the radius 1). Pythagorean theorem leads to the equation (cos 30°)^2 + (sin 30°)^2 = 1. And at the point (cos 30°, sin 30°) the polarization line intersects the unit circle. So you can see a and b, they are the horizontal and vertical coordinates cos 30° and sin 30°.

The numerical values of sin and cos occurring here for the photon polarized at 30° are a = cos 30° = $\sqrt{3}/2 \approx 0.866$, and b = sin 30° = 0.5. Thus, it is the qubit

$$0.866 \cdot |0\rangle + 0.5 \cdot |1\rangle.$$

The situation is analogous for the angle of 120°. Here one can see why a and b can also be negative (their squares are of course always positive). Because for the photon polarized with 120° the numerical values are a = cos 120° = −0.5 and b = sin 120° = $\sqrt{3}/2 \approx 0.866$.

So it's the qubit

$$-0.5 \cdot |0\rangle + 0.866 \cdot |1\rangle.$$

If you don't want to look that closely, or don't want to deal with sines and cosines, just do the math:

$$0.866^2 + 0.5^2 = 1 \left(\text{except for rounding}\right), \text{and} \left(-0.5\right)^2 + 0.866^2 = 1.$$

The graphical representation of a qubit as a straight line in a plane, as in Fig. 8.1, is illustrative. It is a good model for a polarized photon. But it is difficult to extend to a system of multiple qubits.

---

[3]Other physical realizations are based on the spin of ions in ion traps or on superconductivity. In general, all physical objects that behave according to the laws of quantum physics, i.e. in particular that can be entangled, are suitable as quantum bits.

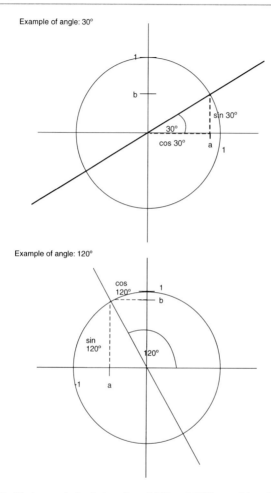

Fig. 8.1 Photons polarized at angles of 30° and 120° as qubits with their probability amplitudes. The resulting probabilities are 1/4 and 3/4

This is because even with only two qubits, it must be possible to distinguish in the graphical representation whether they are entangled or not. So it is not sufficient to simply draw both qubits in two drawings next to each other - here the information about entanglement would not be included.

For the representation of a qubit $|q\rangle = a \cdot |0\rangle + b \cdot |1\rangle$ we therefore choose in this book a new representation, which can be better extended to several qubits, and at the same time one can easily imagine the elementary quantum operations with this representation.

This graphical representation of the qubit $|q\rangle = a \cdot |0\rangle + b \cdot |1\rangle$ consists of two squares with side length 1, which are connected by a line. One of the squares represents the basis state $|0\rangle$, the other the basis state $|1\rangle$. Inside the squares are smaller squares with the area of the probabilities a^2 and b^2, respectively, i.e. with the side lengths of the probability amplitudes a and b, respectively, where the sign of the amplitude determines the color of the inner square. If the amplitude is positive, the inner square is green. If it is negative, it is red.

Figure 8.2 shows the qubit $0.866 \cdot |0\rangle + 0.5 \cdot |1\rangle$ three times. The orientation in space is still irrelevant, it will later determine the position of the qubit in a system.

Figure 8.3 shows three times the qubit $-0.5 \cdot |0\rangle + 0.866 \cdot |1\rangle$. Again, the orientation in space is still irrelevant. But note that the sum of the two small square areas is always 1.

Note for Readers with Knowledge of Complex Numbers This book is an introduction to the basics of quantum computing, a first start. For ease of understanding, we therefore restrict the probability amplitudes to real numbers. This is unusual, and also insufficient for advanced quantum algorithms. In the scientific literature, the probability amplitudes are always complex numbers.

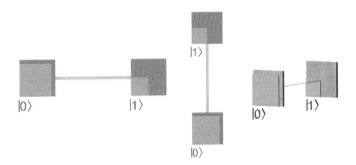

Fig. 8.2 Three different representations of the qubit $0.866 \cdot |0\rangle + 0.5 \cdot |1\rangle$

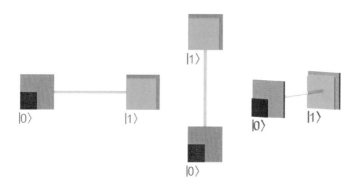

Fig. 8.3 Three different representations of the qubit $-0.5 \cdot |0\rangle + 0.866 \cdot |1\rangle$

Fig. 8.4 Example of extension of the graphical representation to complex amplitudes: The qubit $0.5 \cdot e^{i \cdot 90°} \cdot |0\rangle + 0.866 \cdot e^{-i \cdot 135°} \cdot |1\rangle$

Thus, in advanced literature, a quantum bit is an object of the form

$$|q\rangle = a \cdot |0\rangle + b \cdot |1\rangle,$$

where the probability amplitudes a and b are COMPLEX numbers with $|a|^2 + |b|^2 = 1$. The corresponding probabilities are again the squared absolute values.

The graphical representation of a qubit (and later also of a system of qubits) can be adapted by following Richard Feynman [1] and representing the amplitudes as vectors of the complex number plane, drawing the probabilities as counterclockwise squares, and adding the underlying maximum possible squares with side length 1. The coloring of the small squares can then be omitted; "red" small squares would be recognized by their positioning in the upper right corner. Figure 8.4 shows an example.

On the physical meaning of the complex phases: In the case of photons as an example of qubits, the complex phase of the probability amplitude refers to the phase, a property of the oscillation of the light particle in case of circular or elliptical polarization. This quantity is, of course, physically significant. However, it was not needed in the first part of the book, for the description of the phenomenon of entanglement, and is not needed for the introduction to quantum gates and for teleportation in the second part.

8.2   Quantum Registers Consisting of Two and Three Qubits

In quantum computing, a quantum register $|q_1\,q_2\rangle$ consisting of two qubits is an object of the shape

$$|q_1\,q_2\rangle = c \cdot |00\rangle + d \cdot |01\rangle + e \cdot |10\rangle + f \cdot |11\rangle,$$

that can be measured.

Here $|00\rangle$, $|01\rangle$, $|00\rangle$, $|10\rangle$ and $|11\rangle$ are the so-called "basis states" of the register. They are the states $|0\rangle\,|0\rangle$, $|0\rangle\,|1\rangle$, $|1\rangle\,|0\rangle$ and $|1\rangle\,|1\rangle$ in a shortened notation. The first state always refers to the first qubit, and the second state refers to the second qubit. c, d, e and f are again real numbers between -1 and $+1$ in this book, they are called "probability amplitudes" of the basis states.[4] It must hold $c^2 + d^2 + e^2 + f^2 = 1$.

Here comes an example. Let $c = 1/2$, $d = 1/\sqrt{2}$, $e = -1/\sqrt{8}$ and $f = 1/\sqrt{8}$, so $c = 0.5$, $d \approx 0.707$, $e \approx -0.354$ and $f \approx 0.354$.

Convince yourself that $c^2 + d^2 + e^2 + f^2$ really adds up to 1. So the example quantum register is

$$|q_1q_2\rangle = \frac{1}{2}\cdot|00\rangle + \frac{1}{\sqrt{2}}\cdot|01\rangle - \frac{1}{\sqrt{8}}\cdot|10\rangle + \frac{1}{\sqrt{8}}\cdot|11\rangle$$

$$\approx 0.5\cdot|00\rangle + 0.707\cdot|01\rangle - 0.354\cdot|10\rangle + 0.354\cdot|11\rangle.$$

Hard to imagine? Figure 8.5 shows the graphical representation of the example register:

---

[4]For more advanced applications, c, d, e, and f are complex numbers whose absolute squares add to 1, just as in the case of a single qubit.

Fig. 8.5 The quantum
register
$0.5 \cdot |00\rangle$
$+ 0.707 \cdot |01\rangle$
$- 0.354 \cdot |10\rangle$
$+ 0.354 \cdot |11\rangle$ from two
qubits in the graphical
representation

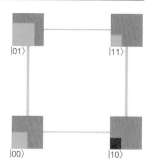

The graphical representation of a quantum register with two qubits

$$c \cdot |00\rangle + d \cdot |01\rangle + e \cdot |10\rangle + f \cdot |11\rangle$$

always consists of four squares with side length 1, which are arranged at the corners of a large square.

The lower left square represents the basis state $|00\rangle$, the upper left square the basis state $|01\rangle$, the lower right square the basis state $|10\rangle$ and the upper right square the basis state $|11\rangle$. So the first qubit is responsible for the first number in the basis state, which is represented in the left-right axis. The second qubit is responsible for the second number in the basis state, which is represented in the bottom-up axis.

Within the squares, as in the graphical representation of a single qubit, there are again smaller squares with the area of the probabilities c^2, d^2, e^2 and f^2 respectively, i.e. with the side lengths of the probability amplitudes c, d, e and f respectively, whereby the sign of the amplitude determines the color of the inner square. If the amplitude is positive, the inner square is green. If it is negative, it is red.

Measuring the register $c \cdot |00\rangle + d \cdot |01\rangle + e \cdot |10\rangle + f \cdot |11\rangle$ gives the following results:

- If one measures the first qubit of the register, one obtains with probability $c^2 + d^2$ the basis state $|0\rangle$ and with probability $e^2 + f^2$ the basis state $|1\rangle$ for this one qubit. Which state the second qubit is then in is dealt with in the next chapter.

- If one measures the second qubit of the register, one obtains
  with probability $c^2 + e^2$ the basis state $|0\rangle$ and with probability
  $d^2 + f^2$ the basis state $|1\rangle$. Which state the first qubit is then in
  is again the subject of the next chapter.
- If we measure both qubits of the register, we get with probabil-
  ity c^2 the basis state $|0\rangle$ for the first qubit and also the basis
  state $|0\rangle$ for the second qubit. For this we briefly write $|00\rangle$ as
  the measurement result.

Similarly, with probability d^2 we obtain the basis state $|0\rangle$ for
the first qubit and the basis state $|1\rangle$ for the second qubit (short-
hand notation $|01\rangle$). With probability e^2 we obtain the basis state
$|1\rangle$ for the first qubit and the basis state $|0\rangle$ for the second qubit
(shorthand notation $|10\rangle$), and with probability f^2 we obtain the
basis state $|1\rangle$ for both qubits (shorthand notation $|11\rangle$).[5]

In the first part of the book, systems of two photons were considered, the first photon was with Alice, the second was with Bob.
In the first experiment, both photons were unentangled. Measuring
each of the four possible outcomes of passing or absorbing the
photon at Alice and Bob gave equal probability 1/4.

As quantum registers these two photons are in the state

$$|q_1 q_2\rangle = \frac{1}{2} \cdot |00\rangle + \frac{1}{2} \cdot |01\rangle + \frac{1}{2} \cdot |10\rangle + \frac{1}{2} \cdot |11\rangle.$$

In the second experiment, both photons were then entangled:
Each one, when measured, gave 1/2 the result of passing or
absorbing with equal probability. But the results of both measurements were always the same. As quantum registers the photons of
the second experiment are in the state

$$|q_1 q_2\rangle = \frac{1}{\sqrt{2}} \cdot |00\rangle + 0 \cdot |01\rangle + 0 \cdot |10\rangle + \frac{1}{\sqrt{2}} \cdot |11\rangle.$$

---

[5]For the computational model of quantum algorithms, there are no further
constraints on the amplitudes c, d, e and f (except that their squares must add
up to 1). Physically, however, it has not yet been possible to realize an arbitrarily specified quantum register of only 2 qubits, let alone of even more
qubits. This is the subject of research in quantum computing hardware. It is
briefly discussed in the last chapter of the book.

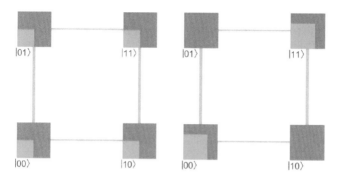

Fig. 8.6 Graphical representation of the photon pairs from experiment 1 (left, unentangled) and experiment 2 (right, entangled)

Figure 8.6 shows the graphical representation of both registers.

In general, if you want to find out whether the two qubits of a given register $c \cdot |00\rangle + d \cdot |01\rangle + e \cdot |10\rangle + f \cdot |11\rangle$ are entangled or unentangled, you can calculate that.

Because if they are unentangled, the first qubit has the shape $a_1 \cdot |0\rangle + b_1 \cdot |1\rangle$ and the second qubit has the shape $a_2 \cdot |0\rangle + b_2 \cdot |1\rangle$. Both can be multiplied, the register then has the shape

$$\left(a_1 \cdot |0\rangle + b_1 \cdot |1\rangle\right) \cdot \left(a_2 \cdot |0\rangle + b_2 \cdot |1\rangle\right) =$$
$$a_1 a_2 \cdot |00\rangle + a_1 b_2 \cdot |01\rangle + a_1 b_2 \cdot |10\rangle + b_1 b_2 \cdot |11\rangle.$$

Thus, if it is possible to find a_1, a_2, b_1, and b_2 with $a_1 a_2 = c$, $a_1 b_2 = d$, $a_2 b_1 = e$, and $a_2 b_2 = f$, the register consists of unentangled qubits. Otherwise it consists of entangled qubits.

Quantum Registers Consisting of Three and More Qubits   The concept of the quantum register with two qubits can be generalized to quantum registers with three qubits (for the purposes of this book) or to quantum registers with an arbitrary number n of qubits (for reading other books on quantum algorithms).

In quantum computing, a quantum register $|q_1 q_2 q_3\rangle$ consisting of three qubits is an object of the shape

$$|q_1 q_2 q_3\rangle = p \cdot |000\rangle + q \cdot |001\rangle + r \cdot |010\rangle + s \cdot |011\rangle$$
$$+ t \cdot |100\rangle + u \cdot |101\rangle + v \cdot |110\rangle + w \cdot |111\rangle$$

that can be measured. Again, the $|.. \rangle$ are the basis states, and the factors p, q, r, s, t, u, v and w their probability amplitudes. The probability amplitudes are again real (complex in the advanced literature) numbers whose squares add up to 1. The first number in a basis state denotes the state of the first qubit in that state, the second number that of the second qubit, and the third number that of the third qubit.

The graphical representation of a quantum register of three qubits consists of eight squares of side length 1 arranged at the corners of a large cube.

The square in the lower left front represents the basis state $|000\rangle$, the square in the lower left back represents the basis state $|001\rangle$, the square in the lower right front represents the basis state $|100\rangle$ and so on. So the first qubit is responsible for the first number in the basis state, which is represented in the left-right axis. The second qubit is responsible for the second number in the basis state, which is represented in the bottom-up axis. And the third qubit is responsible for the third number in the basis state, which is represented in the front-back axis.

Within the squares, as in the graphical representation of a single qubit or in a register of two qubits, there are again smaller squares with the area of the probabilities, i.e. with the side lengths of the probability amplitudes, whereby the sign of the amplitude determines the color of the inner square. If the amplitude is positive, the inner square is green. If it is negative, it is red.

Figure 8.7 shows the example

Fig. 8.7 The quantum register
$0.5 \cdot |000\rangle$
$- 0.612 \cdot |001\rangle$
$- 0.354 \cdot |100\rangle$
$- 0.5 \cdot |111\rangle$ of three qubits in the graphical representation

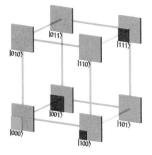

$$|q_1 q_2 q_3\rangle = \frac{1}{2} \cdot |000\rangle - \frac{\sqrt{3}}{\sqrt{8}} \cdot |001\rangle + 0 \cdot |010\rangle + 0 \cdot |011\rangle$$

$$- \frac{1}{\sqrt{8}} \cdot |100\rangle + 0 \cdot |101\rangle + 0 \cdot |110\rangle - \frac{1}{2} \cdot |111\rangle$$

$$\approx 0.5 \cdot |000\rangle - 0.612 \cdot |001\rangle$$

$$- 0.354 \cdot |100\rangle - 0.5 \cdot |111\rangle$$

You can measure a single qubit, two qubits or all three qubits. From the squares of the probability amplitudes, one can add up the probability with which one obtains a certain measurement result.

If, for example, one measures the second qubit, one obtains with probability $p^2 + q^2 + t^2 + u^2$ the value $|0\rangle$. Because this is the sum of the probabilities of the basis states in which the second qubit has the value $|0\rangle$. With probability $r^2 + s^2 + v^2 + w^2$ one obtains the value $|1\rangle$.

If you measure e.g. the first and the second qubit, you get with probability $p^2 + q^2$ the value $|00\rangle$. If you want, you can see for yourself with which probabilities you get the other values:).

The measurement is explained in the graphical representation model in the next subchapter.

In general, a quantum register consisting of an arbitrary number n of qubits is an object of the shape

$$|q_1 q_2 \ldots q_n\rangle = x_0 \cdot |0\ldots00\rangle + x_1 \cdot |0\ldots01\rangle$$
$$+ \ldots + x_{2^n-1} \cdot |11\ldots1\rangle,$$

where $x_0, \ldots, x_{2^n-1}$ again are the probability amplitudes whose squares (if you think complex: magnitude squares) sum up to 1. Graphically, this register can be represented as an *n-dimensional* cube. For $n = 4$ you can still imagine inside - outside as an additional axis, but from $n = 5$ it becomes difficult. Nevertheless, the idea proves to be helpful when considering advanced quantum algorithms.

8.3   Measurement in Quantum Registers

If one measures a single qubit in the state $a \cdot |0\rangle + b \cdot |1\rangle$, then with probability a^2 one obtains the basis state $|0\rangle$, and the qubit is then in this basis state. With probability b^2 one obtains the basis state $|1\rangle$, and the qubit is then in this basis state.

But what happens to a quantum register when you measure one (or more) of the qubits in it?

To answer this question, we first leave the quantum world.

We imagine tossing three coins (namely the first, the second and the third). Each one returns either 0 or 1 as the result. Before looking to the outcome, all we can say about the result is that it is one of the following 8 possibilities:

- $(0, 0, 0), (0, 0, 1), (0, 1, 0), (0, 1, 1), (1, 0, 0), (1, 0, 1), (1, 1, 0),$ or $(1, 1, 1)$.

This world of possibilities forms the **state space of** the experiment. It can be arranged graphically at the corners of a cube, with the first coin determining the left-right position of the respective outcome, the second coin the bottom-up position and the third coin the front-back position, see Fig. 8.8.

If we now "measure" the first of the three coins by looking to see what its outcome is, the state space of the experiment changes. If the outcome of the coin is 0, then we live in a world where the state space consists of the outcomes $(0, 0, 0), (0, 0, 1), (0, 1, 0),$ or $(0, 1, 1)$. If the outcome of the coin is 1, then we live in a world where the state space consists of the outcomes $(1, 0, 0), (1, 0, 1),$

Fig. 8.8 The state space of a toss with three coins, graphically represented as corners of a cube

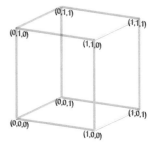

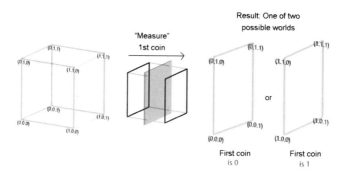

Fig. 8.9   The first of the three coins is looked at. The result is one of two possible worlds (so-called "state spaces")

$(1, 1, 0)$, or $(1, 1, 1)$. Figure 8.9 shows how this can be represented graphically. Measuring removes the edges of the cube that indicate uncertainty with respect to the first coin, that is, the left-right edges. What remains, if the first coin showed 0, are the edges and corners of the left side face of the cube. If the first coin showed 1, the edges and corners of the right side face of the cube remain.

Similarly, when you "measure" (i.e., look up) the second coin, you either get the state space with outcomes $(0, 0, 0)$, $(0, 0, 1)$, $(1, 0, 0)$, or $(1, 0, 1)$ if the outcome of the second coin was 0. Or you get the state space with outcomes $(0, 1, 0)$, $(0, 1, 1)$, $(1, 1, 0)$, or $(1, 1, 1)$ if the outcome of the second coin was 1. Graphically, in the original state space, the edges are removed from the bottom to the top, indicating the uncertainty with respect to the second coin. What remains are the edges and vertices of the bottom and top side faces. Figure 8.10 shows the situation.

Measuring the third coin analogously divides the state space into the two subspaces in which the third coin is 0 and 1, respectively. Graphically, the front-back edges are removed in the cube, showing the uncertainty with respect to the third coin. Figure 8.11 shows how in the graphical representation the edges and corners of the front and back side faces of the original state space remain.

If one "measures" two coins by looking at them, only two results are possible for the system after measurement. For example, if one measures the first and the third coin, either both are 0.

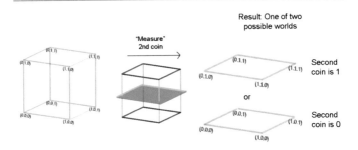

Fig. 8.10 The second of the three coins is looked at. The result is again one of two possible worlds

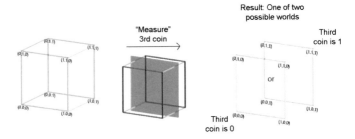

Fig. 8.11 The third of the three coins is looked at. The result is the front or the back "partial world"

Then the total system is in the state $(0, 0, 0)$ or $(0, 1, 0)$. Or the first is 0, the third is 1. Then the total system is in state $(0, 0, 1)$ or $(0, 1, 1)$. Or the first is 1, the third is 0. Then the total system is in state $(1, 0, 0)$ or $(1, 1, 0)$. Or both are 1. Then the total system is in state $(1, 0, 1)$ or $(1, 1, 1)$. So measuring two coins leads to one of four worlds, each of which still contains two possible outcomes: 0 and 1 for the third, unmeasured coin. Graphically, this is represented by removing all edges of the original cube that indicate uncertainty with respect to the first and third coins. These are all left-right edges and all front-back edges. What remains are the edges of the second coin, see Fig. 8.12.

If you look at all three coins, there is no more uncertainty. One of eight possible worlds remains as the state space. Graphically this is represented by removing all edges of the original cube, see Fig. 8.13.

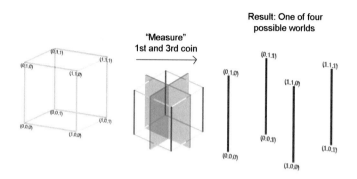

Fig. 8.12   The first and third of the three coins are looked at. Result is one of four possible worlds

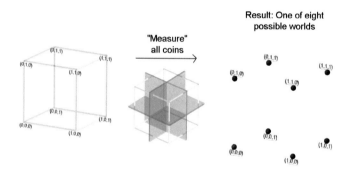

Fig. 8.13   All coins are looked at. Result is one of eight worlds, there is no more uncertainty

This graphical illustration of the measurement process can be applied to the situation where coin flips are not looked up, but where qubits are measured in quantum registers.

The situation here is more complex only insofar as not all eight possible outcomes have the same probability, as was the case with the coin toss. Entangled states of a quantum register cannot be simulated by tossing coins. Not even if the coins are allowed to be unfair, that is, heads and tails do not have the same probability. Another difference between coins and quantum particles is that the coins are in a state when flipped, and that state is preserved when measured. Qubits don't take on a state until they are measured.

Nevertheless, the process of measuring is quite analogous to looking up the result of a coin toss: knowing the result removes uncertainty about the state of the overall register. Graphically, this can be represented in exactly the same way as for coins, by removing the corresponding edges in the state space.

An example illustrates the measurement. The example has no special physical meaning, but was chosen because it is not too complex, but still shows all essential effects. Considered is a quantum register with three qubits in the state

$$| q_1 q_2 q_3 \rangle = \frac{\sqrt{2}}{4} \cdot | 000 \rangle + 0 \cdot | 001 \rangle - \frac{\sqrt{2}}{4} \cdot | 010 \rangle - \frac{\sqrt{3}}{4} \cdot | 011 \rangle$$

$$- \frac{1}{4} \cdot | 100 \rangle + \frac{1}{4} \cdot | 101 \rangle + \frac{1}{2} \cdot | 110 \rangle + \frac{\sqrt{3}}{4} \cdot | 111 \rangle.$$

Measuring the first qubit yields the result $|0\rangle$ with probability

$$\frac{2}{16} + 0 + \frac{2}{16} + \frac{3}{16} = \frac{7}{16}.$$

The quantum register is then in a state that corresponds to the left side of the cube in the graphical representation: the first qubit here is $|0\rangle$. The small squares in this world keep their color. Their size is adjusted so that the sum of the areas of the small squares is 1 again. This is achieved by multiplying the areas by 16/7, i.e. by multiplying the edge lengths by $\sqrt{16/7}$.[6]

Thus, in the case where the first qubit gave the measurement result $|0\rangle$, the quantum register is in the state

$$| q_1 q_2 q_3 \rangle = \sqrt{\frac{16}{7}} \cdot \left(\frac{\sqrt{2}}{4} \cdot | 000 \rangle + 0 \cdot | 001 \rangle \right.$$

$$\left. - \frac{\sqrt{2}}{4} \cdot | 010 \rangle - \frac{\sqrt{3}}{4} \cdot | 011 \rangle \right)$$

$$= \frac{\sqrt{2}}{\sqrt{7}} \cdot | 000 \rangle + 0 \cdot | 001 \rangle - \frac{\sqrt{2}}{\sqrt{7}} \cdot | 010 \rangle - \frac{\sqrt{3}}{\sqrt{7}} \cdot | 011 \rangle.$$

---

[6]It is a computation with conditional probabilities, where each probability comes from a probability amplitude.

With probability 9/16, measuring the first qubit yields the result $|1\rangle$. The quantum register is then in a state which corresponds to the right side face of the cube in the graphical representation.

The state is (after multiplying the probability amplitudes by $\sqrt{16/9}$, to get back to an area sum of 1):

$$|q_1\,q_2\,q_3\rangle = -\frac{1}{3}\cdot|100\rangle + \frac{1}{3}\cdot|101\rangle + \frac{2}{3}\cdot|110\rangle - \frac{\sqrt{3}}{3}\cdot|111\rangle.$$

Figure 8.14 shows the graphical representation.

If one measures in the register of Fig. 8.14 not only the first, but the first and also the third qubit, one obtains (as with the coins) a state space consisting of four possible worlds:

- With probability 1/4 both qubits are $|0\rangle$.
  The register is then in the state $\sqrt{1/2}\cdot|000\rangle - \sqrt{1/2}\cdot|010\rangle$.
- With probability 3/16, the first qubit is $|0\rangle$, and the third is $|1\rangle$.
  The register is then in the state $0\cdot|001\rangle - 1\cdot|011\rangle$.
  (So here there is also no more uncertainty concerning the second qubit, it is always in state $|1\rangle$.)
- With probability 5/16, the first qubit is $|1\rangle$, and the third is $|0\rangle$.
  The register is then in the state $-\sqrt{1/5}\cdot|100\rangle + \sqrt{4/5}\cdot|110\rangle$.
- With probability 1/4 both qubits are $|1\rangle$.
  The register is then in the state $\sqrt{1/4}\cdot|101\rangle + \sqrt{3/4}\cdot|111\rangle$.

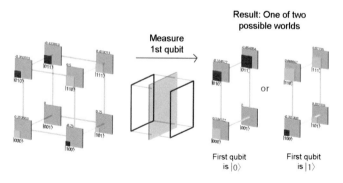

Fig. 8.14 The first of the three qubits is measured. Result is one of two worlds

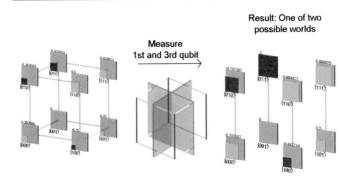

Fig. 8.15 The first and third qubit is measured. There are four possible results

Figure 8.15 shows the graphical representation of the quantum register when the first and third qubit are measured.

If all three qubits are measured, the state of the register is a basis state. There are eight possibilities for this:

- $|000\rangle$, $|001\rangle$, $|010\rangle$, $|011\rangle$, $|100\rangle$, $|101\rangle$, $|110\rangle$ and $|111\rangle$.

Reference

1. R. Feynman. *QED: Die seltsame Theorie des Lichts und der Materie.* Piper Verlag GmbH, May 2018.

Quantum Gates on One Qubit

9

Classical computer science is based on circuits. In these, so called Boolean gates are applied to bits in a predetermined order. The best known Boolean gates are: The NOT gate (applied to a bit, turns the value 0 into 1 and vice versa), and the AND, OR and XOR gates, which are applied to two bits.

The term "algorithm" is much more abstract. Someone who develops a computer game, for example, usually does not think in terms of circuits. Nevertheless, circuits form the basis, the hardware, for all classical computers.

Circuits also form the basis for quantum algorithms. These are quantum circuits in which quantum gates are applied to qubits. The most important gates on a qubit are the (Pauli)-X gate, the (Pauli)-Z gate and the Hadamard gate. They are introduced in the following and get along with real numbers.[1]

---

[1]There are other gates on a qubit whose descriptions require the complex numbers. The best known of these is the (Pauli)-Y gate.

© Springer-Verlag GmbH Germany, part of Springer Nature 2022   69
B. Just, *Quantum Computing Compact*,
https://doi.org/10.1007/978-3-662-65008-0_9

9.1   Pauli-X, Pauli-Z and Hadamard (X, Z and H): Gates on One Qubit

The Pauli-X gate (short X) and the Pauli-Z gate (short Z) are named after the physicist Wolfgang Pauli, the Hadamard gate (short H) is named after the mathematician Jacques Hadamard.

It is not easy to see what operations one can apply to a photon without destroying its polarization. But it turns out that one can mirror the polarization at certain simple angles.[2] This is exactly what Pauli-X, Pauli-Z and H do:

* Pauli-X reflects at the 45°-degree-angle,
* Pauli-Z reflects at the 0°-degree-angle, and
* H reflects at the 22.5°-degree-angle, i.e. at half the angle of Pauli-X.

Figure 9.1 shows the new polarization that results when the gates are applied to a photon polarized at 120°, that is, to $-0.5 \cdot |0\rangle + \sqrt{3/4} \cdot |1\rangle$.

You can also calculate the new polarization instead of drawing it. Here come the calculation rules, which apply not only to photons, but to all qubits.

Applied to a qubit $a \cdot |0\rangle + b \cdot |1\rangle$, Pauli-X, Pauli-Z and H give the following results:

---

[2]Some remarks on the technical realization with photons:

* Photons can (theoretically) be mirrored at highly reflecting mirrors without their entanglement properties being significantly affected or cancelled. Mirroring with such mirrors is therefore not a "measurement" in the quantum physical sense.
* The reflection at such a mirror does not reflect the polarization of a qubit, so it is not a realization of Pauli-X, Pauli- Z or Hadamard. The reflection only changes the flight direction of a photon.
* In order to technically realize Pauli-X, Pauli-Z and Hadamard for photons, beam splitters and waveplates (and highly reflecting mirrors) are used as components. This is technically demanding and not possible without errors. Here, too, care must be taken that the manipulation of the photons is not a physical measurement process which would influence the entanglement properties.

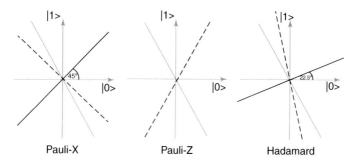

Fig. 9.1 Application of the quantum gates X, Z and H to a photon polarized at 120° angle

$$\text{Pauli} - X : b \cdot |0\rangle + a \cdot |1\rangle$$
$$\text{Pauli} - Z : a \cdot |0\rangle - b \cdot |1\rangle$$
$$\text{Hadamard} : \frac{a+b}{\sqrt{2}} \cdot |0\rangle + \frac{a-b}{\sqrt{2}} \cdot |1\rangle$$

Applied to the basis state |0⟩, Pauli-X thus returns the basis state |1⟩. Pauli-Z returns the state |0⟩, and Hadamard returns the state $\sqrt{1/2} \cdot |0\rangle + \sqrt{1/2} \cdot |1\rangle$. Hadamard is the "shaking" from Chap. 7. Hadamard transforms a probability of 1 (i.e., a surely predictable measurement result, a full square in the graphical representation) to two probabilities of 1/2.

Applied to the basis state |1⟩, Pauli-X yields the basis state |0⟩. Paul-Z yields the state -|1⟩, and Hadamard yields the state $\sqrt{1/2} \cdot |0\rangle - \sqrt{1/2} \cdot |1\rangle$.

Applied to the example qubit, $-0.5 \cdot |0\rangle + \sqrt{3/4} \cdot |1\rangle$ the gates yield the following results:

$$\text{Pauli} - X : \sqrt{3/4} \cdot |0\rangle - 0.5 \cdot |1\rangle$$
$$\text{Pauli} - Z : -0.5 \cdot |0\rangle - \sqrt{3/4} \cdot |1\rangle$$
$$\text{Hadamard} : \frac{-1+\sqrt{3}}{\sqrt{8}} \cdot |0\rangle + \frac{-1-\sqrt{3}}{\sqrt{8}} \cdot |1\rangle$$

Figure 9.2 shows the example in the graphical representation:

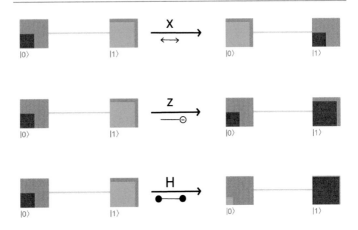

Fig. 9.2 The X, Z and H gates are applied to the qubit $-0.5 \cdot |0\rangle + \sqrt{3/4} \cdot |1\rangle$

9.2   X, Z, H, Applied to a Qubit in a Quantum Register

X, Z and H are quantum gates that are always applied to a single qubit. However, if this qubit is located in a quantum register consisting of several qubits, its change alters the state of the entire register.

How a register of two or three qubits changes, if X, Z or H is applied to one of the qubits in it, is the subject of the following considerations. On the one hand it is always calculated how the probability amplitudes of the basic states change, and on the other hand it is shown how the change can be graphically represented.

First, the change of a register consisting of two qubits is considered when Pauli-X is applied to the first of the two qubits. If the register is in a basis state, Pauli-X returns a basis state again. This leaves the second qubit unchanged and makes the first 1 if it was 0, and vice versa. So by Pauli-X applied to the first of the two qubits you get

- from the state $|00\rangle$ the state $|10\rangle$;
- from the state $|01\rangle$ the state $|11\rangle$;
- from the state $|10\rangle$ the state $|00\rangle$;
- from the state $|11\rangle$ the state $|01\rangle$.

States which are not basis states are called "mixed states". If the register is in a mixed state, a mixed state is obtained by Pauli-X. This swaps the amplitudes of the basis states accordingly.

See the example of the register in Fig. 8.5:

$$|q_1 q_2\rangle = \frac{1}{2} \cdot |00\rangle + \frac{1}{\sqrt{2}} \cdot |01\rangle - \frac{1}{\sqrt{8}} \cdot |10\rangle + \frac{1}{\sqrt{8}} \cdot |11\rangle$$

$$\approx 0.5 \cdot |00\rangle + 0.707 \cdot |01\rangle - 0.354 \cdot |10\rangle + 0.354 \cdot |11\rangle.$$

Application of Pauli-X to the first qubit provides as new register state

$$|q_1 q_2\rangle = \frac{1}{2} \cdot |10\rangle + \frac{1}{\sqrt{2}} \cdot |11\rangle - \frac{1}{\sqrt{8}} \cdot |00\rangle + \frac{1}{\sqrt{8}} \cdot |01\rangle$$

$$\approx 0.5 \cdot |10\rangle + 0.707 \cdot |11\rangle - 0.354 \cdot |00\rangle + 0.354 \cdot |01\rangle,$$

i.e. in the usual grouping of the basic states, by rearranging the summands

$$|q_1 q_2\rangle = -\frac{1}{\sqrt{8}} \cdot |00\rangle + \frac{1}{\sqrt{8}} \cdot |01\rangle + \frac{1}{2} \cdot |10\rangle + \frac{1}{\sqrt{8}} \cdot |11\rangle$$

$$\approx -0.354 \cdot |00\rangle + 0.354 \cdot |01\rangle + 0.5 \cdot |10\rangle + 0.707 \cdot |11\rangle,$$

In the graphical representation, the new state is obtained by swapping the amplitudes along the two horizontal edges in the quadrilateral. This is because the first qubit is responsible for the left-right axis in the graphical representation, see Fig. 9.3:

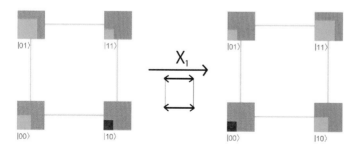

Fig. 9.3 Pauli-X is applied to the first qubit of the quantum register $0.5 \cdot |00\rangle + 0.707 \cdot |01\rangle - 0.354 \cdot |10\rangle + 0.354 \cdot |11\rangle$: Graphical representation

If Pauli-X is applied to the second qubit of a register consisting of two qubits, then for the basis states you obtain

- from the state $|00\rangle$ the state $|01\rangle$;
- from the state $|01\rangle$ the state $|00\rangle$;
- from the state $|10\rangle$ the state $|11\rangle$;
- from the state $|11\rangle$ the state $|10\rangle$.

If the register is in a mixed state, a mixed state is again obtained by Pauli-X as in the application to the first qubit. This swaps the amplitudes of the basis states in the graphical representation along the up-down edges, because the second qubit is responsible for these.

Figure 9.4 illustrates the application of Pauli-X to the second qubit in the example from above.

Mathematically, from the original state:

$$|q_1 q_2\rangle = \frac{1}{2} \cdot |00\rangle + \frac{1}{\sqrt{2}} \cdot |01\rangle - \frac{1}{\sqrt{8}} \cdot |10\rangle + \frac{1}{\sqrt{8}} \cdot |11\rangle$$

$$\approx 0.5 \cdot |00\rangle + 0.707 \cdot |01\rangle - 0.354 \cdot |10\rangle + 0.354 \cdot |11\rangle.$$

by applying Pauli-X to the second qubit one obtains the new register state

$$|q_1 q_2\rangle = \frac{1}{2} \cdot |01\rangle + \frac{1}{\sqrt{2}} \cdot |00\rangle - \frac{1}{\sqrt{8}} \cdot |11\rangle + \frac{1}{\sqrt{8}} \cdot |10\rangle$$

$$\approx 0.5 \cdot |01\rangle + 0.707 \cdot |00\rangle - 0.354 \cdot |11\rangle + 0.354 \cdot |10\rangle,$$

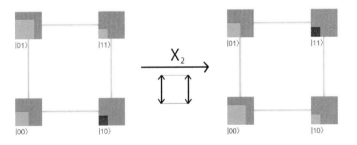

Fig. 9.4 Pauli-X is applied to the second qubit of the quantum register $0.5 \cdot |00\rangle + 0.707 \cdot |01\rangle - 0.354 \cdot |10\rangle + 0.354 \cdot |11\rangle$: Graphical representation

i.e. in the usual grouping of the basis states, by rearranging the summands

$$|q_1 q_2\rangle = \frac{1}{\sqrt{2}} \cdot |00\rangle + \frac{1}{2} \cdot |01\rangle + \frac{1}{\sqrt{8}} \cdot |10\rangle - \frac{1}{\sqrt{8}} \cdot |11\rangle$$

$$\approx 0.707 \cdot |00\rangle + 0.5 \cdot |01\rangle + 0.354 \cdot |10\rangle - 0.354 \cdot |11\rangle.$$

If Pauli-X is applied to a qubit in a register of three qubits, this also leads to swapping of the probability amplitudes. In the graphical representation, it is easy to see which amplitudes are swapped. If Pauli-X is applied to the first qubit in the register, the amplitudes along the four edges in the left-right direction are swapped, because that is what the first qubit is responsible for. Applying Pauli-X to the second qubit swaps along the four bottom-up edges, and applying Pauli-X to the third qubit swaps along the four front-back edges.

As an example, a Pauli X gate is applied to the third qubit of the quantum register in Fig. 8.7. The register is in the state

$$|q_1 q_2 q_3\rangle = \frac{1}{2} \cdot |000\rangle - \frac{\sqrt{3}}{\sqrt{8}} \cdot |001\rangle + 0 \cdot |010\rangle + 0 \cdot |011\rangle$$

$$- \frac{1}{\sqrt{8}} \cdot |100\rangle + 0 \cdot |101\rangle + 0 \cdot |110\rangle - \frac{1}{2} \cdot |111\rangle$$

$$\approx 0.5 \cdot |000\rangle - 0.612 \cdot |001\rangle$$

$$- 0.354 \cdot |100\rangle - 0.5 \cdot |111\rangle.$$

By applying Pauli-X to the third qubit, the amplitudes of the basis states which differ exactly at the third digit, are swapped in each case.

You get

$$|q_1 q_2 q_3\rangle = \frac{1}{2} \cdot |001\rangle - \frac{\sqrt{3}}{\sqrt{8}} \cdot |000\rangle + 0 \cdot |011\rangle + 0 \cdot |010\rangle$$

$$- \frac{1}{\sqrt{8}} \cdot |101\rangle + 0 \cdot |100\rangle + 0 \cdot |111\rangle - \frac{1}{2} \cdot |110\rangle$$

$$\approx 0.5 \cdot |001\rangle - 0.612 \cdot |000\rangle$$

$$- 0.354 \cdot |101\rangle - 0.5 \cdot |110\rangle.$$

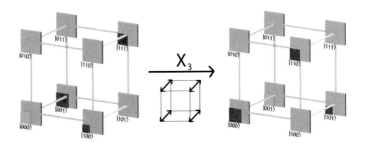

Fig. 9.5 In the quantum register $0.5 \cdot |000\rangle - 0.612 \cdot |001\rangle - 0.354 \cdot |100\rangle - 0.5 \cdot |111\rangle$ Pauli-X is applied to the third qubit

The graphical representation from Fig. 9.5 is clearer.

Pauli-X, applied to a qubit in a register of multiple qubits, thus swaps the probability amplitudes along the edges for which the qubit is responsible in the graphical representation.

Now consider how the Hadamard transform H acts on a register when applied to one of the qubits in the register. As a reminder: If H is applied to a single qubit, then it changes

- the basis state $|0\rangle$ the state $\dfrac{1}{\sqrt{2}} \cdot |0\rangle + \dfrac{1}{\sqrt{2}} \cdot |1\rangle$, which is approximately the state $0.707 \cdot |0\rangle + 0.707 \cdot |1\rangle$, and
- the basis state $|1\rangle$ the state $\dfrac{1}{\sqrt{2}} \cdot |0\rangle - \dfrac{1}{\sqrt{2}} \cdot |1\rangle$, which is approximately the state $0.707 \cdot |0\rangle - 0.707 \cdot |1\rangle$.

Figure 9.6 shows this in te graphical representation.

If the qubit is not in a basis state, the two effects overlap, as in the example in Fig. 9.2 at the beginning of the chapter.[3]

If H is applied to a (single) qubit in a register consisting of two or three qubits, the overall state of the register changes, just like when Pauli-X is applied. In the graphical representation, just as with Pauli-X, we consider the edges for which the qubit is responsible, i.e., the left-right edges for the first qubit, the down-up edges for the second qubit, and the front-back edges for the third

---

[3]If you want, you can do the math: Applying the Hadamard transformation twice returns the original state. In the realization with photons this is clear: Mirroring twice always returns the original state.

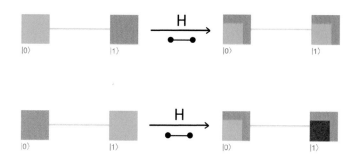

Fig. 9.6 The Hadamard transform H is applied to a single qubit in a basis state. In the transition in the upper row the basis state is $|0\rangle$, in the transition in the lower row it is $|1\rangle$

qubit. Along these edges, the probability amplitudes are "smeared" in H, while they were swapped in Pauli-X.

Figure 7.3 from the chapter "Quantum algorithms graphically" already showed in the graphical representation the change of a quantum register of three qubits, if on the initial state $|000\rangle$ the Hadamard transformation H is applied first to the first, then to the second and then to the third qubit. Now we consider the example when H is applied to a register of three qubits first to the first, then to the second, and then to the third qubit, and the register is in basis state $|001\rangle$ at the beginning. The following transitions are obtained mathematically, whereby in each case one summand becomes two summands after a Hadamard transformation (which are each in a line here):

$$|001\rangle \xrightarrow{H_1} \frac{1}{\sqrt{2}} \cdot |001\rangle + \frac{1}{\sqrt{2}} \cdot |101\rangle$$

$$\xrightarrow{H_2} \frac{1}{\sqrt{2}} \cdot \frac{1}{\sqrt{2}} \cdot |001\rangle + \frac{1}{\sqrt{2}} \cdot \frac{1}{\sqrt{2}} \cdot |011\rangle$$

$$+ \frac{1}{\sqrt{2}} \cdot \frac{1}{\sqrt{2}} \cdot |101\rangle + \frac{1}{\sqrt{2}} \cdot \frac{1}{\sqrt{2}} \cdot |111\rangle$$

$$\xrightarrow{H_3} \frac{1}{\sqrt{2}} \cdot \frac{1}{\sqrt{2}} \cdot \frac{1}{\sqrt{2}} \cdot |000\rangle - \frac{1}{\sqrt{2}} \cdot \frac{1}{\sqrt{2}} \cdot \frac{1}{\sqrt{2}} \cdot |001\rangle$$

$$+ \frac{1}{\sqrt{2}} \cdot \frac{1}{\sqrt{2}} \cdot \frac{1}{\sqrt{2}} \cdot |011\rangle - \frac{1}{\sqrt{2}} \cdot \frac{1}{\sqrt{2}} \cdot \frac{1}{\sqrt{2}} \cdot |011\rangle$$

$$+ \frac{1}{\sqrt{2}} \cdot \frac{1}{\sqrt{2}} \cdot \frac{1}{\sqrt{2}} \cdot |100\rangle - \frac{1}{\sqrt{2}} \cdot \frac{1}{\sqrt{2}} \cdot \frac{1}{\sqrt{2}} \cdot |101\rangle$$

$$+ \frac{1}{\sqrt{2}} \cdot \frac{1}{\sqrt{2}} \cdot \frac{1}{\sqrt{2}} \cdot |110\rangle - \frac{1}{\sqrt{2}} \cdot \frac{1}{\sqrt{2}} \cdot \frac{1}{\sqrt{2}} \cdot |111\rangle,$$

which is approximately the result

$$\approx 0.354 \cdot \Big(|000\rangle - |001\rangle + |010\rangle - |011\rangle$$

$$+ |100\rangle - |101\rangle + |110\rangle - |111\rangle \Big).$$

In the graphical representation in Fig. 9.7 the changes can more easily be tracked.

At the end of the chapter, we will see how Pauli-Z, applied to a single qubit in a quantum register of several qubits, changes the state of the register.

Pauli-Z, applied to a single qubit $a|0\rangle + b|1\rangle$, changes the sign of the probability amplitude of $|1\rangle$. The result is therefore $a|0\rangle - b|1\rangle$. In graphical terms, this means that a green square at state $|1\rangle$ becomes a red square of the same size, a red square at state $|1\rangle$ becomes a green square of the same size, and everything else remains unchanged. An example can be found in the middle of Fig. 9.2.

Thus, if Pauli-Z is applied to a qubit in a quantum register of two or three qubits, the probability amplitudes of those basis

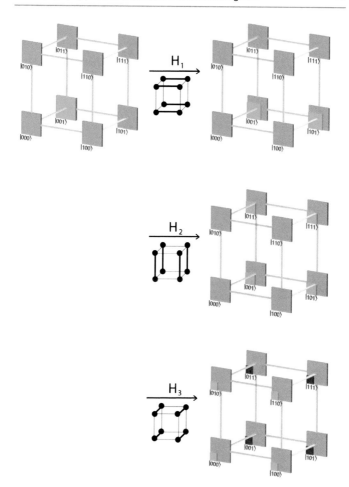

Fig. 9.7 On a quantum register in the basis state $|001\rangle$, the Hadamard transform is applied first to the first, then to the second, and then to the third qubit

states change their sign (their color in the graphical representation) where this qubit is $|1\rangle$.

Three examples illustrate this complicated proposition.

(i) Pauli-Z is applied to the first qubit of the quantum register

$$|q_1 q_2\rangle = \frac{1}{2} \cdot |00\rangle - \frac{1}{\sqrt{2}} \cdot |01\rangle - \frac{1}{\sqrt{8}} \cdot |10\rangle + \frac{1}{\sqrt{8}} \cdot |11\rangle$$
$$\approx 0.5 \cdot |00\rangle + 0.707 \cdot |01\rangle - 0.354 \cdot |10\rangle + 0.354 \cdot |11\rangle.$$

The result is

$$|q_1 q_2\rangle = \frac{1}{2} \cdot |00\rangle + \frac{1}{\sqrt{2}} \cdot |01\rangle + \frac{1}{\sqrt{8}} \cdot |10\rangle - \frac{1}{\sqrt{8}} \cdot |11\rangle$$
$$\approx 0.5 \cdot |00\rangle + 0.707 \cdot |01\rangle + 0.354 \cdot |10\rangle - 0.354 \cdot |11\rangle.$$

Figure 9.8 shows the graphical representation.

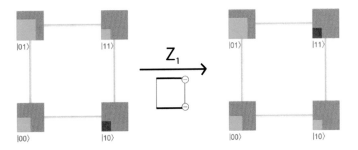

Fig. 9.8 Pauli-Z is applied to the first qubit of the quantum register $0.5 \cdot |00\rangle + 0.707 \cdot |01\rangle - 0.354 \cdot |10\rangle + 0.354 \cdot |11\rangle$: Graphical representation

(ii) Pauli-Z is applied to the second qubit of the quantum register

$$|q_1 q_2\rangle = \frac{1}{2} \cdot |00\rangle + \frac{1}{\sqrt{2}} \cdot |01\rangle - \frac{1}{\sqrt{8}} \cdot |10\rangle + \frac{1}{\sqrt{8}} \cdot |11\rangle$$

$$\approx 0.5 \cdot |00\rangle + 0.707 \cdot |01\rangle - 0.354 \cdot |10\rangle + 0.354 \cdot |11\rangle.$$

The result is

$$|q_1 q_2\rangle = \frac{1}{2} \cdot |00\rangle - \frac{1}{\sqrt{2}} \cdot |01\rangle - \frac{1}{\sqrt{8}} \cdot |10\rangle - \frac{1}{\sqrt{8}} \cdot |11\rangle$$

$$\approx 0.5 \cdot |00\rangle - 0.707 \cdot |01\rangle - 0.354 \cdot |10\rangle - 0.354 \cdot |11\rangle.$$

Figure 9.9 shows the graphical representation.

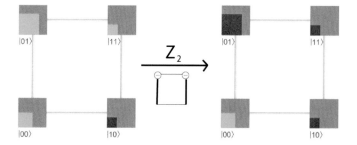

Fig. 9.9 Pauli-Z is applied to the second qubit of the quantum register 0.5 · |00⟩ + 0.707 · |01⟩ − 0.354 · |10 ⟩ + 0.354 · |11⟩: Graphical representation

(iii) Pauli-Z is applied to the third qubit of the quantum register

$$|q_1 q_2 q_3\rangle = \frac{1}{2} \cdot |000\rangle - \frac{\sqrt{3}}{\sqrt{8}} \cdot |001\rangle - 0 \cdot |010\rangle + 0 \cdot |011\rangle$$

$$- \frac{1}{\sqrt{8}} \cdot |100\rangle + 0 \cdot |101\rangle + 0 \cdot |110\rangle - \frac{1}{2} \cdot |111\rangle.$$

$$\approx 0.5 \cdot |000\rangle - 0.612 \cdot |001\rangle$$

$$- 0.354 \cdot |100\rangle - 0.5 \cdot |111\rangle.$$

The result is

$$|q_1 q_2 q_3\rangle = \frac{1}{2} \cdot |000\rangle + \frac{\sqrt{3}}{\sqrt{8}} \cdot |001\rangle + 0 \cdot |010\rangle - 0 \cdot |011\rangle$$

$$- \frac{1}{\sqrt{8}} \cdot |100\rangle - 0 \cdot |101\rangle + 0 \cdot |110\rangle + \frac{1}{2} \cdot |111\rangle.$$

$$\approx 0.5 \cdot |000\rangle + 0.612 \cdot |001\rangle$$

$$- 0.354 \cdot |100\rangle + 0.5 \cdot |111\rangle.$$

Figure 9.10 shows the graphical representation.

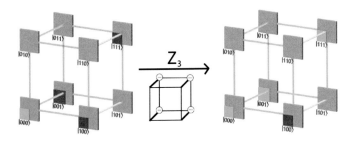

Fig. 9.10 In the quantum register $0.5 \cdot |000\rangle - 0.612 \cdot |001\rangle - 0.354 \cdot |100\rangle$ $- 0.5 \cdot |111\rangle$ Pauli-Z is applied to the third qubit

CNOT: A Quantum Gate on Two Qubits

10

CNOT (controlled NOT), i.e. "controlled negation", is a quantum gate that requires two qubits as input: A control bit, and a target bit. The way it works, very roughly speaking, is that as much as the control bit is in the $|1\rangle$ state, the target bit is negated, i.e. its amplitudes are swapped. This may sound a bit cryptic at the moment and will be explained in more detail below.

If the quantum register consists of two qubits, CNOT can therefore be applied in two ways: Either with the first qubit as the control bit and the second as the target bit, or vice versa.

If the quantum register consists of three qubits, it can be applied in six different ways. If you want, you can count them.

10.1 CNOT in a Register Consisting of Two Qubits

The negation of a classical bit consists in setting its state "0" when it was previously "1", and vice versa.

The negation of a qubit is the Pauli-X: its state becomes $b \cdot |0\rangle + a \cdot |1\rangle$ if it was previously $a \cdot |0\rangle + b \cdot |1\rangle$, and vice versa. Controlled negation CNOT in quantum computing can be most closely compared to conditional negation in classical computing: If the condition is satisfied, the negation takes place, otherwise it does not. CNOT applies Pauli-X to the target bit when the control

© Springer-Verlag GmbH Germany, part of Springer Nature 2022
B. Just, *Quantum Computing Compact*,
https://doi.org/10.1007/978-3-662-65008-0_10

bit is $|1\rangle$. If the control bit is $|0\rangle$, Pauli-X is not applied to the target bit. And if the control bit is neither all $|0\rangle$ nor all $|1\rangle$, but has a mixed state? Then Pauli-X is applied to the target bit "as much" as the control bit was $|1\rangle$.

For a system $|q_1\,q_2\rangle$ of two qubits, this means when CNOT is applied with the first qubit as the control bit and the second qubit as the target bit[1]:

- If the system is in state $|00\rangle$, it is still in state $|00\rangle$ after the CNOT.
- If the system is in state $|01\rangle$, it is still in state $|01\rangle$ after the CNOT.
- If the system is in state $|10\rangle$, it is in state $|11\rangle$ after the CNOT.
- If the system is in state $|11\rangle$, it is in state $|10\rangle$ after the CNOT.
- If the system is in the state $c \cdot |00\rangle + d \cdot |01\rangle + e \cdot |10\rangle + f \cdot |11\rangle$, then after applying CNOT it is in the state $c \cdot |00\rangle + d \cdot |01\rangle + e \cdot |11\rangle + f \cdot |10\rangle$, i.e. in the state
  $c \cdot |00\rangle + d \cdot |01\rangle + f \cdot |10\rangle + e \cdot |11\rangle$.

Here is an example in a register consisting of two qubits, which is in the following state:

$$|q_1\,q_2\rangle = -\sqrt{0.2} \cdot |00\rangle + \sqrt{0.3} \cdot |01\rangle + \sqrt{0.1} \cdot |10\rangle - \sqrt{0.4} \cdot |11\rangle$$
$$\approx -0.447 \cdot |00\rangle + 0.548 \cdot |01\rangle + 0.316 \cdot |10\rangle - 0.633 \cdot |11\rangle.$$

---

[1] The technical realization of a CNOT gate is difficult for photons (and also for other realizations of qubits), but indispensable for most quantum algorithms. How can one photon control another?

The basic procedure is always as follows: The control bit is brought into an "environment" and generates a resonance there. The "environment" can be, for example, an atom trapped in a cage, whereby the cage for photons can be constructed, for example, from mirrors and laser beams. The target bit is then brought into the environment, which is in a certain oscillation state due to the contact with the control bit. This oscillation state then causes the change in the target bit. See, for example, [1, 2].

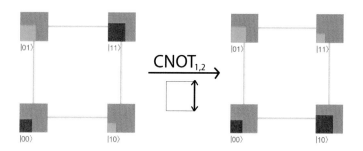

Fig. 10.1 A CNOT is applied in a register consisting of two qubits. Control bit is the first, target bit the second qubit

If CNOT is applied with the first qubit as the control bit, and the second qubit as the target bit, we get:

$$| q_1 q_2 \rangle = -\sqrt{0.2} \cdot | 00 \rangle + \sqrt{0.3} \cdot | 01 \rangle - \sqrt{0.4} \cdot | 10 \rangle + \sqrt{0.1} \cdot | 11 \rangle$$
$$\approx -0.447 \cdot | 00 \rangle + 0.548 \cdot | 01 \rangle - 0.633 \cdot | 10 \rangle + 0.316 \cdot | 11 \rangle.$$

Figure 10.1 shows the graphical representation. In principle, the amplitudes are swapped along up-down edges, because the second qubit (the target bit) is responsible for these. But not along both edges the amplitudes are swapped, but only at the basis states, where the target bit, i.e. the first qubit, is $|1\rangle$. Thus only at the right bottom-up edge of the square.

If, with the same input register state

$$| q_1 q_2 \rangle = -\sqrt{0.2} \cdot | 00 \rangle + \sqrt{0.3} \cdot | 01 \rangle + \sqrt{0.1} \cdot | 10 \rangle - \sqrt{0.4} \cdot | 11 \rangle$$
$$\approx -0.447 \cdot | 00 \rangle + 0.548 \cdot | 01 \rangle + 0.316 \cdot | 10 \rangle - 0.633 \cdot | 11 \rangle$$

CNOT is applied, where now the second qubit is the control bit and the first qubit is the target bit, you get

$$| q_1 q_2 \rangle = -\sqrt{0.2} \cdot | 00 \rangle - \sqrt{0.4} \cdot | 01 \rangle + \sqrt{0.1} \cdot | 10 \rangle + \sqrt{0.3} \cdot | 11 \rangle$$
$$\approx -0.447 \cdot | 00 \rangle - 0.633 \cdot | 01 \rangle + 0.316 \cdot | 10 \rangle + 0.584 \cdot | 11 \rangle.$$

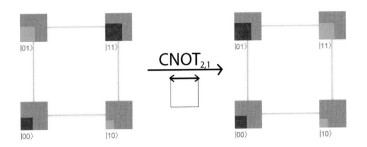

Fig. 10.2 A CNOT is applied in a register consisting of two qubits. Control bit is the second qubit, the target bit is the first qubit

Figure 10.2 shows the graphical representation. The amplitudes are swapped along left-right edges, because the first qubit is responsible for these. The second qubit controls along which edges are swapped: Only basis states where the second qubit is 1 are swapped. This is the upper left-right edge of the square.

10.2   CNOT in a Register Consisting of Three Qubits

An example now illustrates a CNOT when applied to two qubits in a register of three qubits. Consider a quantum register with three qubits in state

$$|q_1 q_2 q_3\rangle = \frac{\sqrt{2}}{4} \cdot |000\rangle + 0 \cdot |001\rangle - \frac{\sqrt{2}}{4} \cdot |011\rangle - \frac{\sqrt{3}}{4} \cdot |011\rangle$$

$$- \frac{1}{4} \cdot |100\rangle + \frac{1}{4} \cdot |101\rangle + \frac{1}{2} \cdot |110\rangle + \frac{\sqrt{3}}{4} \cdot |111\rangle.$$

A CNOT is applied to this with the second qubit as the control bit and the first qubit as the target bit. This causes Pauli-X to be applied to those basis states of the first qubit (the target bit) where the second qubit is in the $|1\rangle$ state. That is, amplitudes of states $|010\rangle$ and $|110\rangle$ are swapped with each other, amplitudes of states

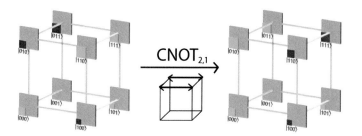

Fig. 10.3 A CNOT is applied in a register consisting of three qubits. Control bit is the second, target bit the first qubit

$|011\rangle$ and $|111\rangle$ are swapped with each other. After the CNOT, the register is in state

$$| q_1 q_2 q_3 \rangle = \frac{\sqrt{2}}{4} \cdot | 000 \rangle + 0 \cdot | 001 \rangle + \frac{1}{2} \cdot | 010 \rangle + \frac{\sqrt{3}}{4} \cdot | 011 \rangle$$

$$- \frac{1}{4} \cdot | 100 \rangle + \frac{1}{4} \cdot | 101 \rangle - \frac{\sqrt{2}}{4} \cdot | 110 \rangle - \frac{\sqrt{3}}{4} \cdot | 111 \rangle.$$

Figure 10.3 shows the graphical representation. The amplitudes are swapped along left-right edges, because the first qubit is responsible for these. But not all four edges are swapped. The second qubit controls along which edges are swapped: Only at the basis states, where the second qubit is $|1\rangle$. These are the two upper left-right edges of the cube.

References

1. Wei-Bo Gao, Ping Xu, Xing-Can Yao, Otfried Gühne, Adán Cabello, Chao-Yang Lu, Cheng-Zhi Peng, Zeng-Bing Chen, and Jian-Wei Pan. Experimental realization of a controlled-not gate with four-photon six-qubit cluster states. *Phys. Rev. Lett.*, 104:020501, Jan 2010.
2. Bastian Hacker, Stephan Welte, Gerhard Rempe, and Stephan Ritter. A photon-photon quantum gate based on a single atom in an optical resonator. *Nature*, 536(7615):193–196, July 2016.

Teleportation

11

Teleportation is one of the most spectacular algorithms in quantum information science. But it is not matter that is teleported here, but information.[1] Teleportation is suitable for transmitting information quickly over long distances in a vacuum, e.g. between satellites.

In this chapter you will find the algorithm step by step. But first here is the rough setting, and the structure of the algorithm, based on photons:

At two different places in space, at Alice's and at Bob's, there is one photon each from an entangled photon pair. This photon pair was previously produced and one of its photons was sent to Alice, the other one to Bob, who now store it at their place, e.g. in glass fiber cables.

Alice has a third photon. The goal is to transfer the polarization of this photon (not the photon itself) to Bob's photon.

This is done by Alice cleverly applying the Hadamard transform and CNOT to her photon and the third photon. Then she measures her photon and the third photon. This causes Bob's photon to change its properties, according to the rules of quantum physics, via the "spooky action at a distance".

Bob's photon stores then all properties to the former polarization of the third photon, but it is not yet polarized itself in the

---

[1] So quantum physics can't do a "beam me up, Scotty".

© Springer-Verlag GmbH Germany, part of Springer Nature 2022
B. Just, *Quantum Computing Compact*,
https://doi.org/10.1007/978-3-662-65008-0_11

exact way the third photon previously was. For this, Bob still has to apply Pauli-X and/or Pauli-Z to his photon in a suitable way. What exactly he has to do, Alice tells him on a classical channel (e.g.: by phone). It depends on her measurement result, and that is random. Only after Bob has applied these suitable Pauli transformations, his photon is polarized in the same way as Alice's third photon was before.

The observant reader notes:

(i) Bob's photon is not instantaneously, i.e. in particular not with superluminal speed, transferred into the state of the third photon. This holds, since Alice has to tell Bob her measurement result, and this can be done at most with the speed of light.

(ii) Neither Alice nor Bob know the polarization of the third photon that is transmitted. They can't measure it either, because that would change it. They only know that Bob's photon has this polarization after the teleportation.

11.1   The Algorithm for Teleportation

The **start state** for teleportation consists of three qubits. Alice and Bob each have one qubit of an entangled photon pair. In addition, Alice has another qubit in the state $a \cdot |0\rangle + b \cdot |1\rangle$. However, she does not know the state itself, so she has no information about a and b.

Alice's and Bob's qubit are thus in state $\sqrt{1/2} \cdot |00\rangle + \sqrt{1/2} \cdot |11\rangle$, see Fig. 8.6 from Chap. 8.

Figure 11.1 now shows the starting state when the third qubit is in the $0.5 \cdot |0\rangle - \sqrt{3/4} \cdot |1\rangle$ state (unknown to Alice and Bob).

This example is used to illustrate the steps of teleportation in this section. You can see from it how the information about a and b travels through the cube.

The three qubits are first considered as a quantum register with three qubits. It is in the state

$$\frac{1}{\sqrt{2}} \cdot a \cdot |000\rangle + \frac{1}{\sqrt{2}} \cdot b \cdot |001\rangle + \frac{1}{\sqrt{2}} \cdot a \cdot |110\rangle + \frac{1}{\sqrt{2}} \cdot b \cdot |111\rangle.$$

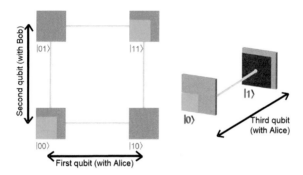

Fig. 11.1 The initial state for teleportation: Alice and Bob each have one of two entangled qubits, plus one qubit at Alice. She does not know its state, for the graphical representation it is exemplary $0.5 \cdot |0\rangle - \sqrt{3/4} \cdot |1\rangle$

Fig. 11.2 The start state for teleportation: The register from Fig. 11.1 seen as ONE register consisting of three qubits

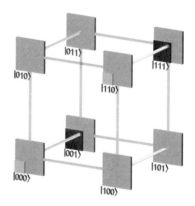

Figure 11.2 shows the starting state overall for values $a = 0.5$, and $b = -\sqrt{3/4}$, i.e. $b \approx -0.866$.

Step 1 Alice applies CNOT with the third qubit as the control bit and her own, the first, qubit as the target bit. This causes the quantum register with the three qubits to enter the state.

$$\frac{1}{\sqrt{2}} \cdot a \cdot |000\rangle + \frac{1}{\sqrt{2}} \cdot b \cdot |011\rangle + \frac{1}{\sqrt{2}} \cdot b \cdot |101\rangle + \frac{1}{\sqrt{2}} \cdot a \cdot |110\rangle.$$

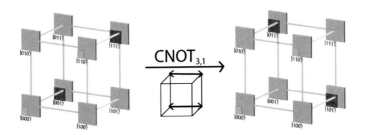

Fig. 11.3 Step 1: Alice applies CNOT with the third qubit as control bit and her own (the first) qubit as target bit

Figure 11.3 shows the transformation again for $a = 0.5$, and $b = -\sqrt{3/4}$.

Step 2 Alice applies the Hadamard transform to the third qubit. If you want, you can calculate that afterwards the register is in the following state:

$$\frac{1}{2} \cdot a \cdot |000\rangle + \frac{1}{2} \cdot a \cdot |001\rangle + \frac{1}{2} \cdot b \cdot |010\rangle + \frac{1}{2} \cdot (-b) \cdot |011\rangle$$

$$+ \frac{1}{2} \cdot b \cdot |100\rangle + \frac{1}{2} \cdot (-b) \cdot |101\rangle + \frac{1}{2} \cdot a \cdot |110\rangle + \frac{1}{2} \cdot a \cdot |111\rangle.$$

Figure 11.4 shows the step for the example from above with $a = 0.5$, and $b = -\sqrt{3/4}$.

Step 3 Alice measures the two qubits that are with her. In the register, these are qubit 1 and qubit 3. Each can be in state $|0\rangle$ or state $|1\rangle$, so there are four possible measurement results. Alice knows which state Bob's qubit is in depending on the measurement result. You can see this clearly in the graphical representation. Alice's measurements remove all left-right edges (for which the first qubit is responsible), and front-back edges (for which the third qubit is responsible). What remains is one of four worlds for the second qubit (at Bob), see Fig. 11.5 for the example $a = 0.5$, and $b = -\sqrt{3/4}$.

For the sake of completeness, here are the calculated values for Alice's knowledge of Bob's qubit:

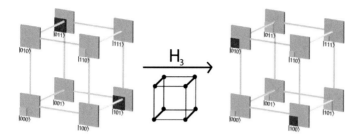

Fig. 11.4 Step 2: Alice applies the Hadamard transform to the third qubit

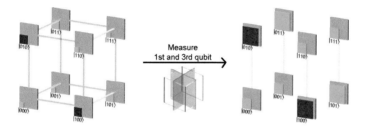

Fig. 11.5 Step 3: Alice measures the first and third qubit

- If both of her measurement results are $|0\rangle$, Bob's qubit is in the state $a \cdot |0\rangle + b \cdot |1\rangle$.
- If both of her measurement results are $|1\rangle$, Bob's qubit is in the state $-b \cdot |0\rangle + a \cdot |1\rangle$.
- If the measurement result of her qubit is $|0\rangle$, and that of the third qubit is $|1\rangle$, Bob's qubit is in the state $a \cdot |0\rangle - b \cdot |1\rangle$.
- If the measurement result of her qubit is $|1\rangle$, and that of the third qubit is $|0\rangle$, then Bob's qubit is in the state $b \cdot |0\rangle + a \cdot |1\rangle$.

If you want, you can calculate it with the help of the rules for measuring qubits, see Sect. 8.3 about measuring in quantum registers. However, the graphical representation is certainly more descriptive.

Step 4 Alice calls Bob and tells him the result of her measurement. Bob now also knows which world he is in.

Step 5 Bob brings his qubit to the original state $a \cdot |0\rangle + b \cdot |1\rangle$ of the third qubit based on the information from Alice:

- If both of Alice's measurement results were $|0\rangle$, Bob's qubit is already in the state $a \cdot |0\rangle + b \cdot |1\rangle$. He then does nothing.
- If both of Alice's measurement results were $|1\rangle$, Bob's qubit is in the state $- b \cdot |0\rangle + a \cdot |1\rangle$. He first applies Pauli-X, and then Pauli-Z, and gets $a \cdot |0\rangle + b \cdot |1\rangle$.
- If the measurement result of Alice's qubit was $|0\rangle$, and that of the third qubit was $|1\rangle$, Bob's qubit is in the state $a \cdot |0\rangle - b \cdot |1\rangle$. He applies Pauli-Z to bring it to the state $a \cdot |0\rangle + b \cdot |1\rangle$.
- If the measurement result of Alice's qubit was $|1\rangle$, and that of the third qubit was $|0\rangle$, Bob's qubit is in the state $b \cdot |0\rangle + a \cdot |1\rangle$. He applies Pauli-X and brings it to the state $a \cdot |0\rangle + b \cdot |1\rangle$.

For the example $a = 0.5$, and $b = -\sqrt{3/4}$, Fig. 11.6 shows Bob's actions.

Bob's qubit is now in the state that the third qubit was originally in at Alice.

So, theoretically, quantum teleportation is possible. The question is whether it can also be carried out in practice.

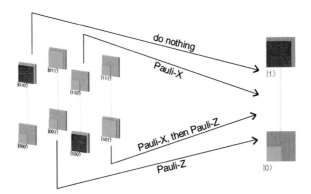

Fig. 11.6 Step 5: Depending on Alice's measurement result, by applying Pauli X and / or Pauli-Z Bob brings his qubit into the original state of the single (third) qubit located with Alice

11.2   State of Practical Implementation

The quantum teleportation algorithm was found as a theoretical possibility in 1993 by a group of American, Canadian, French and Israeli researchers [1]. Over the years, ever greater distances were then bridged experimentally via teleportation of photons. Here comes a list of milestones.

- In 1997, a research group led by Zeilinger in Vienna succeeded for the first time in demonstrating teleportation in the laboratory;
- In 2003, a research group led by Gisin succeeded in teleporting over 55 m in Geneva, for the first time outside the laboratory;
- In 2004, Zeilinger's group succeeded in bridging a distance of 600 m, from one bank of the Danube to the other;
- In 2010, a Chinese research group in Shanghai led by Xian Min Lin succeeded in bridging a distance of 16 km;
- In 2012, a research group at the Chinese University of Science and Technology led by Pan Jian-Wei bridged a distance of 97 km;
- Also in 2012, the working group around Zeilinger bridged the distance of 143 km between La Palma and Tenerife;
- In 2017, an international working group, including Zeilinger and Pan Jian-Wei, bridged the distance of 1400 km from Earth to the Chinese quantum satellite Micius;
- The working group also bridged a distance of 7600 km between Austria and China in 2017.

There have also been successful experiments on the teleportation of properties of other objects, i.e. not photons, but atoms, for example. Teleportation of more complex properties of so-called "QTrits", a kind of pair of qubits, has also been performed.

The bridging of large distances with quantum entanglement or quantum teleportation is interesting in the context of a "quantum internet" to be built, which is currently being researched intensively. Communication between spatially distant participants in this internet could run via the instantaneous, wireless "entanglement channel".

However, very small distances are also important (and are currently being researched more intensively than large distances). This is because chips within a quantum computer should, if possible, be able to communicate with each other via the "entanglement channel" even without a material connection. For small distances, it is not so important that the communication is fast, but above all that it also takes place without energy consumption and thus without heating.

Reference

1. Charles H. Bennett, Gilles Brassard, Claude Crépeau, Richard Jozsa, Asher Peres, and William K. Wootters. Teleporting an unknown quantum state via dual classical and Einstein-Podolsky-Rosen channels. *Phys. Rev. Lett.*, 70:1895–1899, Mar 1993.

Further Quantum Algorithms and Hardware

12

12.1 Further Quantum Algorithms

In 1982 Richard Feynmann dealt with the question whether all physical processes can be simulated on classical computers [4]. It turns out that this is not possible, because classical computers cannot generate real randomness. But quantum physics contains real randomness. So a computer was developed (at first in thought) based on quantum bits, and following the laws of quantum physics. In 1984, with a paper by David Deutsch, it was finally established how a quantum computer are supposed to work [3].

Since then, research has been developed in two directions:

- Hardware: How can quantum computers be built? More on this in Sect. 12.2.
- Software: If you had quantum computers, what problems could you solve (quickly) with them, and what do quantum algorithms look like?

The bridge between the two is - as with classical computers - the circuit. Circuits are the smallest hardware control unit of a computer. And an algorithm consists of a mental series of circuits, one for each input quantity. It's helpful to think of it in terms of the adder: The algorithm for addition using carrying of digits works in principle for numbers with any number of digits, but to

© Springer-Verlag GmbH Germany, part of Springer Nature 2022
B. Just, *Quantum Computing Compact*,
https://doi.org/10.1007/978-3-662-65008-0_12

actually perform addition on a machine, you need a circuit (with finitely many inputs).

The first quantum algorithms were quantum circuits like teleportation presented in the last section. Other quantum circuits were presented for communication or key exchange in cryptography provable secure against eavesdropping [1, 2]. The research area was considered a somewhat exotic discipline pursued by quantum physicists with an interest in computer science.

This changed abruptly when Peter Shor presented a fast quantum algorithm solving the factorization problem in 1994 [7]. Because with the solution of the factorization problem it was clear: A quantum computer, once built, would break a large part of the Internet encryption.[1]

Shor's algorithm is a real algorithm, i.e. mentally a whole sequence of circuits: For each number of digits of a number to be factorized, it returns one circuit (as in the classical adder).

Since Shor's algorithm, a whole series of quantum algorithms have been developed; if you want, google "Quantum Algorithm Zoo". Some of them are considerably faster than classical algorithms. What is the idea behind these algorithms? What makes quantum circuits so much faster than classical circuits?

Quantum circuits can "tune in" to detect hidden structures in the problem.[2]

Imagine a normal chessboard with 8 × 8 squares, and a game figure that starts in one corner and moves straight ahead, square by square. When it reaches the end of a row, it turns around, and

---

[1]The factorization problem is the inverse of multiplication. Multiplication is easy. There should be no difficulty in working out, say, 5153*6397. Written multiplication or a small calculator gives the result 32,963,741. Now in factorization problem, the product is given and the factors are to be found. Up to now, no classical fast algorithm is known for this. Neither in writing nor with a small calculator one can easily find out how to factorize e.g. 34,903,109 (for curious readers: the answer is 4967*7027). The RSA encryption system, which is mostly used on the Internet, uses the fact that factorization problem is difficult.

[2]Mathematicians nobly call such structures "hidden subgroups in the solution space".

continues the next row (in the opposite direction). In this way, it walks the entire playing field.

This figure "notices" at some point that there is a regular structure that has to do with the number 8 (although it itself does not see the chessboard from above). A human being who would have the task of the game piece would quite quickly "swing in" to the 8-rows (the reader may imagine).

If the chessboard looks different, for example, has 65 squares arranged as a 5 × 13 square, the figure would "swing in" very quickly (depending on the direction of travel) to the 5 or the 13. But if the number of squares is a prime number, say 67 squares, no "swinging in" can take place. The chessboard here would consist of a single long row with 67 squares.

"Swinging in" is a natural property of complex systems of all kinds in nature, but also of social systems. Self-organizing systems strive for equilibrium, and often achieve this via some kind of oscillation. Here are some examples:

- A string of a musical instrument that is excited goes into harmonic vibration and then back to rest.
- Chemical reactions of all kinds, once excited, strive for a state of lowest energy.
- Social systems (e.g., social networks) settle into a broadly stable structure of tightly-knit peer groups and a few connections between those peer groups.
- Economic models of markets assume that markets oscillate towards an equilibrium state.
- Traffic flows organise themselves (if not regulated) towards an optimal flow.
- Neural networks in artificial intelligence oscillate towards a (at least local) optimum.

All these processes take place very quickly without external control.

In this sense, a quantum circuit is also a self-organizing system. This is because the quantum bits must not (as in the classical case) be seen only as individual bits, but they influence each other systemically via quantum entanglement. Therefore, quantum algorithms can be used to quickly simulate the behavior of complex systems. Most easily, of course, they can simulate the behavior of a quantum computer itself (which a classical computer has great difficulty doing). But by cleverly encoding the input, they can also simulate other systems. This is of particular interest, for example, in experimental design, or in the development of new materials or drugs. The ability of quantum algorithms to detect hidden structures is also of great importance in the analysis of large amounts of data of all kinds.[3]

There are now many applications for quantum algorithms in civil society use, even beyond cryptography. There are also already good tools for programming quantum algorithms. Any beginner can sign up online at IBM without any prior knowledge, and can play with self-created quantum circuits there. Advanced developers will find Q#, a quantum programming language on a development environment like C#.

The biggest problems in the realization of quantum computers currently lie in the hardware, because it is a long way from a quantum circuit on paper to a quantum circuit that actually exists physically.

12.2   Hardware

Fast quantum algorithms require quantum computers with about a few hundred quantum bits, which can all be entangled with each other, which can be controlled, and which are reasonably stable.[4] However, chips for such quantum computers do not currently

---

[3]At first glance, quantum computers cannot do anything that classical computers cannot do - they are just faster. It requires some effort to construct a problem that classical computers cannot solve at all, but quantum computers can. Here it is used that quantum computers can realize true randomness. So problems in the area of checking randomness can be solved qualitatively better than classical computers [6].

[4]Very roughly, n quantum bits can simulate 2^n classic bits. So 10 quantum bits simulate 1024 classical bits, 20 quantum bits simulate about a million classical bits, and and 40 quantum bits simulate a petabyte.

exist, and their technical feasibility is not foreseeable at the moment. Some researchers believe that sooner or later such chips will be created, others believe this will never be the case.

We are currently in the NISQ era.

"NISQ" is the "Noisy Intermediate-Scale Quantum" technology [5]. The largest quantum chips currently in existence are the Google Bristlecone (72 QBits, from 2018), the Google Sycamore (54 QBits, from 2019), and the Intel Tangle Lake (49 QBits, from 2018) and the IBM Eagle (127 QBits, from 2021). But these are just individual chips, not an entire quantum computer. An important complete quantum computer already sold as a whole computer is the one of IBM. It is a cube with 2 m edge length.[5]

The problem with all currently available quantum chips is their high susceptibility to errors. Quantum bits must not interact with anything, otherwise it's like a measurement for them that changes their state. This is very difficult to realize in practice. The main physical techniques for realizing quantum registers on quantum chips are:

- Photons. These are mainly used in the field of quantum communication and can thus be an essential building block for a future quantum internet. Photons have the advantage of covering distances very quickly (almost at the speed of light) in fibre optic cables or in the vacuum of space between satellites. So far, however, it is beyond any technical reach to entangle more than about 10 photons with each other.
- Superconductivity. The Google, IBM and Intel chips mentioned above are based on small superconducting loops. For

[5]The company D-Wave from Canada produces "quantum computers" with several thousand "quantum bits". The concept here, however, is not the concept of entangled quantum particles, but another concept, the so-called "adiabatic" quantum computer. Here, a problem is encoded into a state of a physical system, which then transforms itself into the ground state of lowest energy when it is cooled down. This ground state corresponds to the solution of the problem. Certainly this is a very interesting concept for approximating optimization problems, but not quantum computing in the true sense (sometimes D-Wave is accused of sailing under a false flag, since they call their computer "quantum computer").

superconductivity to work, the temperature must be close to absolute zero. Extreme cooling must therefore be ensured. In the chips that exist today, moreover, by no means all quantum bits are entangled with each other, but only some that are spatially adjacent to each other on the chip. The susceptibility to errors is also unfortunately high. There are methods for error correction, but these require eight additional QBits per QBit to be corrected. The 72 QBits on the Google bristlecone thus correspond to one byte of corrected QBits.

- Trapped ions. These are ions, i.e. atoms or electrons, which are trapped in a strong magnetic field. Here, too, extreme cooling is needed up to today, but there is a theoretical chance that one day it will no longer be necessary. Researchers in Innsbruck have entangled 20 trapped ions. This may be better than larger chips if there is less chance of error.

Quantum computing in the NISQ era works by interconnecting some (error-prone) NISQ chips via classical computing techniques. The practically implemented algorithms are thus hybrids - partly they work on quantum chips, partly on conventional technology. If necessary, quantum circuits with a few QBits are also cleverly simulated on classical computers.

Google announced Quantum Supremacy in 2019, but the announcement is controversial.

"Quantum Supremacy" means that a problem can be computed faster by a real existing quantum computer than by any existing classical computer. The problem Google cites as proof is very specific to quantum computers, however. Basically, it is a simulation of a (error-prone) quantum computer. In a chat there was a rant about the setting, telling a coffee cup is superior to a quantum computer according to this logic. Because it could simulate faster than a quantum computer what exactly happens when a coffee cup falls off the table and shatters on the floor.

The physics underlying quantum computing, with its "spooky action at a distance", is fascinating. So is the way quantum algorithms work. The hardware is a challenge for physicists, and is being researched internationally at great expense. The applications of fast quantum algorithms are very diverse. How long will the NISQ era last? Will true, undisputed superiority of quantum computers over classical computers ever be achieved? These are questions that the future will answer. But for physicists, mathematicians, computer scientists, philosophers, social theorists, material scientists, logisticians and data analysts, quantum computing is already a highly topical field.

References

1. C. H. Bennett and G. Brassard. Quantum cryptography: Public key distribution and coin tossing. In *Proceedings of IEEE International Conference on Computers, Systems, and Signal Processing*, page 175, India, 1984.

2. Charles H. Bennett, François Bessette, Gilles Brassard, Louis Salvail, and John Smolin. Experimental quantum cryptography. *Journal of Cryptology*, 5(1):3–28, January 1992.

3. David Deutsch. Quantum theory, the Church–Turing principle and the universal quantum computer. *Proceedings of the Royal Society of London. A. Mathematical and Physical Sciences*, 400(1818):97–117, July 1985.

4. Richard P. Feynman. Simulating physics with computers. *International Journal of Theoretical Physics*, 21(6–7):467–488, Jun 1982.

5. John Preskill. Quantum Computing in the NISQ era and beyond. *Quantum*, 2:79, August 2018.

6. Ran Raz and Avishay Tal. Oracle separation of BQP and PH. In Moses Charikar and Edith Cohen, editors, *Proceedings of the 51st Annual ACM SIGACT Symposium on Theory of Computing, STOC 2019, Phoenix, AZ, USA, June 23-26, 2019*, pages 13–23. ACM, 2019.

7. P. W. Shor. Algorithms for quantum computation: discrete logarithms and factoring. In *Proceedings 35th Annual Symposium on Foundations of Computer Science*, pages 124–134, 1994.

Printed in the United States
by Baker & Taylor Publisher Services